Cryptozoology

and the
Investigation of
Lesser-Known
Mystery Animals

Cryptozoology

and the Investigation of Lesser-Known Mystery Animals

Edited by
Chad Arment

Coachwhip Publications
Landisville, Pennsylvania

Contents

Introduction   7

Spiders that Shine   14

The Suiformes: The Expanding World of
    Pigs and Peccaries   21

Coelacanths of Silver: A Biogeography of
    Latimeria?   32

A Mysterious Kentucky Water Cryptid   54

Didi of the Imagination   59

Freshwater Cephalopods   71

Preliminary Notes on a North American
    "Flying Snake"   94

Crypto New Mexico   103

Paul Gauguin's Mystery Bird   115

"Dinosaur" Sightings in the United States   137

Miscellaneous Menagerie   169

About the Authors   223

Introduction

Chad Arment

Mystery animals known by relatively few sightings are saddled with the same caveats as traditional cryptids, but these cautions require greater consideration. While each cryptid must be examined independently from each other (whether or not Bigfoot exists has no relevance to whether a luminous spider exists), the lesser-known cryptids generally share some traits that should be recognized by the investigator as research develops.

With fewer reports, there is a greater possibility that the general description is inaccurate. Few laypersons are capable of reliable identification of uncommon species (particularly cryptic predators and other species that rarely enter a suburban backyard), let alone an unrecognized species. Memory may play tricks, as they attempt to describe an unknown animal using characteristics of well-known creatures. (Thus a head may be wolf-like, horse-like, camel-like, sheep-like, snake-like, etc.) Outright misidentification of familiar animals should always be one of the first possibilities to consider. Some people are just apt, whether due to environmental conditions or unreliable identification skills, to mangle a description and yet be utterly convinced that that is what they saw.

Hoaxing is also an increased possibility. An old newspaper account describing a creature unlike any other in cryptozoology, and which has never been reported since, must be examined with a critical eye. There have also been cases where someone attempted to foist an imaginary creation on cryptozoology. There is a natural tension here, as there may be good reasons for a lack of sighting reports (remote habitat, or a regional culture lacking consistent zoological examination from qualified sources). It isn't critical thinking merely to dislike a report, and so call it a probable hoax. Yet, cryptozoology is not journalism—it is not our job to report facts just to let the reader decide the issue based on personal opinions or feelings. We should be carefully noting and evaluating the sources, the descriptions, the locations, the witnesses, and other data gleaned from the accounts. Obvious difficulties shouldn't be ignored by propping up meticulously detailed scenarios. At the same time, there will be many cases where a solid answer one way or the other is unavailable. This is where illogical debunking usually comes into play—i.e., if there is no unequivocal evidence, the report must be fallacious. Clearly, that is nonsense. Ambiguous reports still serve a purpose, pointing to regions worth exploring, or suggesting potential avenues for investigation—they just shouldn't be used in an attempt to "prove" that an animal exists.

A third issue with lesser-known mystery animals is that the lack of broad public recognition requires different approaches to soliciting reports. Again, there is a dichotomy. It may be harder to find people with pertinent information, requiring greater effort within a targeted region. At the same time, depending on the cryptid, there may be far less stigma in discussing the animal. If, for example, I talked to a group of veteran trappers here in Pennsylvania, they would probably be more willing to discuss the possibility of long-tailed wildcats (previously noted in *Cryptozoology: Science & Speculation*) than Bigfoot.

So how might we investigate a cryptid which, perhaps, is brand new to cryptozoology? First, there is the matter of identifying the mystery animal as a legitimate cryptid. Just because we run across a report of an animal that doesn't immediately resemble a known species, does not automatically qualify it as a cryptid.

There has to be an initial filter for such reports, weeding out mistaken or unreliable accounts. Few investigators adequately provide this filter on their own. Unique reports should be discussed with others who may have experience or knowledge with different aspects of natural history. Once a report passes through this initial filter, and the animal can feasibly be considered a cryptid, it still must continue under critical evaluation—the initial filter provides an educated first response rather than just accepting each and every report that comes along. The filter should be handled with care, of course—I know one investigator who clumsily "identified" an animal using one of a number of purported traits, ignoring more substantial characters, all the while claiming it was "critical thinking" that led his theorizing. That unfortunately models a poor example for those new to cryptozoology.

Next, we start to build a profile of the cryptid. A cryptid is an ethnozoological category, not a taxonomic category, so we are using possible traits as reported by witnesses or others who may have heard about the animal. These include morphological traits (what the animal looks like) and environmental traits (where and under what conditions has it been seen). Because we are not simply observing and measuring a living recognized species, the profile will never be completely accurate. As research develops, however, the more opportunity there is to build a more plausible description. Note, of course, that reports do not build on each other to add greater weight to the theory—that would be a fallacy of numbers. Rather, a greater number of reports offer more data to create a better profile with the intent of developing a way to acquire true confirmative evidence.

Once a profile has been created, this should be made available to other investigators. One of the biggest problems in cryptozoology is that many good reports have disappeared because investigators retired or died (or societies folded) without providing for their files to be made available in the future. Obviously, an active investigation may require privacy, but suitable venues for publication should be sought for end-stage activities. Not every investigator (few, if any) will follow an investigation from initial recognition to obtaining confirmative evidence. Publication ensures that potential avenues for research will stay open for those who have the necessary resources.

Finally, the investigation should focus on using profile data to create a field methodology for the purpose of acquiring evidence that can be tested and used to confirm or disconfirm the true identity of the species. More often than not, investigators get stuck on building a profile, gathering more and more sighting reports. That serves a purpose, but it will never replace confirmative evidence. Too many times, investigators look at sighting reports, photos, and other non-confirmative evidence as if they are piling up on a scale—as if, at some point, the weight of circumstantial evidence tips, and as much as proves the animal exists. Instead, we should consider this sort of evidence as part of a folkloric "radar" screen. Each sighting creates a blip, with the possibility of false readings, but also forming clumps of regularities, giving direction for future efforts.

Of course, discussing investigation begs the question—where and how do we find reports of lesser-known or as-yet unrecognized mystery animals? The first places to look, surprisingly for some, are published accounts by scientists. Browsing through *Deep-Sea Photography* (Hersey 1967), we come across a chapter by Marshall and Bourne that incidentally illustrates several marine animals unidentified at that time. One, a small galeoid shark from the Red Sea, was commonly photographed but never captured. Another photo shows "curious, shiny 'marbles' common in our Gulf of Aden pictures.... These unidentified animals leave tracks in the sediment." These species may have been described since then, but if not, they represent legitimate cryptids.

Regional faunal reviews may also provide clues to potential unknown species. Carey (1972) noted that an early herpetological survey of Anegada island (BVI) by Robert Schomburgk listed "a small red snake, similar in appearance to the coral snake. To my knowledge, there are no snakes in the Greater Puerto Rican area, including Anegada, which resemble in the remotest way coral snakes. It is not known to what creature he was referring."

Specialist newsletters occasionally note potential new species. Matt Bille provides some good examples in his chapter on mystery pigs.

Scientists outside zoology are another potential source. Anthropologists, in particular, and others who examine how cultures relate to animals around them (ethnozoology), may describe reported species but not have the proper background or interest in systematically evaluating those reports. Valeri (2000) noted the presence of the ihisiupu, or "lord of the cuscuses," within Huaulu ethnozoology on the island of Seram in the Moluccas, but never saw an actual specimen. He stated, "I was told that it looks like a rat, that 'it climbs in the trees' (ia saa hoto aitufhu), and that it resembles in its color a species of cuscus (the elahuwa, *Phalanger orientalis orientalis*) which is gray or whitish, but that it is much smaller than that cuscus. Also, contrary to this or any other cuscus, it has no marsupium." Valeri suggested it might be a forest rat or an arboreal civet that preys on cuscuses, and described the associations in Huaulu folklore between animal "masters" or "lords" and the "normal" species they are supposed to rule. The ihisiupu may or may not be a legitimate cryptid— further evaluation would need to be made, both in terms of examining the known mammalian fauna of Seram and attempting to get more details from anyone who has spent time with the Huaulu people. Louis Dupree, an archaeologist, once ran into stories of a giant bird, "a veritable roc," that was supposedly stealing grain from villagers in south-central Afghanistan. Agreeing to stay on guard in wait for the culprit, Dupree eventually discovered that the thief was not a bird at all, but a sch'goon: later identified as an Indian crested porcupine, which hadn't previously been known from that country.

Historically, animal collectors for zoos and museums would have been a good source for strange stories. Unfortunately, very few of them (at least the English-speaking ones) ever wrote down their experiences for future generations. Charles Cordier (1983), in a brief article, collected for the New York Zoological Society and other groups. He stated, "During that first stay in the Congo, while driving along the pot-holed road in the remote region of Lubutu I heard a tremendous screaming and screeching on my right from within the forest. Some 50 or 60 metres in front, I suddenly saw a gorilla emerge from the greenery and cross the road. It was erect. Then I knew nothing about gorillas; later, during our second stay, I actually saw a large creature whose belly was not voluminous nor was its arm overlong as in the gorilla." This is an area where non-English literature could provide potential reports—German, Dutch, Belgian, and French zoo collectors in Africa and other parts of the world may have written accounts worth pursuing.

Historical societies, research libraries, and university archives frequently house regional folklore compilations and newspapers on microfilm to explore. Old science journals published the occasional gem. Historians and local folklore researchers can sometimes provide leads on potential subjects. Interviews with cultural groups (whether grouped by locality or occupation, as with fishermen or oil-rig divers) have been fruitful for many investigators.

Pinpointing regions where unknown species may lurk, and so possibly may be encountered by local peoples, is another possible methodology. This is one area where scientists have done a good bit of work in recent years, spurred on by discoveries of larger species, though they tend to focus on the habitat aspects rather than the ethnoknown possibilities. John MacKinnon (2000) suggests that prime habitat for unrecognized mammals includes those areas with:

1) Long geological stability
2) Tropical richness
3) Long-term humid conditions
4) Poor exploration
5) Semi-isolation or containing habitat islands
6) Relatively small size of any such habitat islands
7) High endemism in non-mammalian species

MacKinnon points out that the North Annam range of Vietnam, where multiple significant new and rediscovered species have been found, matches all of these criteria. I suspect that other habitat characteristics will prove useful in cryptozoology, as habitat-specificity is a common trait among rare species (Brooks 1998).

In the chapters that follow, investigators introduce a number of lesser-known mystery animals. We will see areas where new cryptids might be located, descriptions of mystery animals new to cryptozoology, and even details on purported accounts that turned out to be false or mistaken. There are opportunities for even the beginning cryptozoological investigator, and I hope that this anthology stimulates interest to look beyond the high-profile cryptids to the vast zoological treasury that occasionally nudges itself into our cultural consciousness, waiting to be discovered.

References:

Brooks, Daniel M. 1998. Habitat variability as a predictor of rarity in large Chacoan mammals. *Vida Silvestre Neotropical* 7(2-3): 115-120.

Carey, W. Michael. 1972. The herpetology of Anegada, British Virgin Islands. *Caribbean Journal of Science* 12(1-2): 79-89.

Cordier, Charles. 1983. Reminiscences of live animal collecting in Zaire. *International Zoo News* 178: 26-28.

Dupree, Louis. 1955. Who saw the Sch'goon? *Natural History* (Dec.): 525-527.

Hersey, John Brackett, editor. 1967. *Deep-Sea Photography.* Johns Hopkins Oceanographic Studies No. 3. Baltimore, MD: Johns Hopkins Press.

MacKinnon, John. 2000. New mammals in the 21st century? *Annals of the Missouri Botanical Gardens* 87: 63-66.

Valeri, Valerio. 2000. *The Forest of Taboos.* Madison, WI: University of Wisconsin Press.

Spiders that Shine

Chad Arment

Bioluminescence is common in certain animal groups: deep-sea fish, marine invertebrates, fireflies and other beetles. There are other zoological groupings where we certainly don't expect to find it—spiders, for example. Yet, in a trio of brief notes published in the journal *Science*, two separate sightings of a glowing spider were described. The first was reported by the American Museum of Natural History's energetic curator and fossil hunter, Barnum Brown. Brown traveled to Burma (now Myanmar) in 1923 to hunt fossils in the Pondaung Hills. The following story comes from that trip.

A Luminous Spider

One day in Central Burma the trail in the jungle was exceptionally difficult. It was long past noon when I realized that the return journey would be equally long and tiring.

Camp lay on the other side of a long range of hills and there was a short cut from the main trail that would save several miles, but this trail was faint. I reached the supposed cut-off about dusk and followed it upward. Darkness came on swiftly and my pony began to stumble. Somewhere we had missed the trail, for at intervals I could still glimpse the crest of the hills and I knew my general direction.

Fireflies sparkled here and there. Presently a few feet away I saw a ball of light as large as one's thumb. It was stationary. Tying the horse, I approached it as carefully as possible, finding it surrounded by thorny bushes. It did not move and I pressed the brush aside until I was directly over it and then struck a match. There in full view was a spider, his large oval abdomen grayish, with darker markings. Still he did not move, and as the match died out his abdomen again glowed to full power, a completely oval light, similar in quality to that of the fireflies. Remembering native tales of poisonous insects, I wrapped a handkerchief around one hand, parted the brush with the other, and when close enough made a quick grab. Alas! The handkerchief caught on a stick before I could encircle him and my treasure scurried away. I followed as quickly as possible, but the light soon disappeared under stones, brush or in some burrow, for I never saw it again.

Many nights I searched in the jungle and questioned natives and white officers who had passed through that district, but apparently no one else had reported a luminous spider, nor can I find record of any known elsewhere.

Burmese never leave their houses after dark on account of their fear of spirits, so it is not surprising that the natives had never seen one, but some other traveler may be so fortunate as to capture one of these spiders.

The place where I saw the specimen was between the villages of Kyawdaw and Thitkydaing, Pakkoku

District, about one hundred and twenty miles west of Mandalay, Burma, in April, 1923.

Barnum Brown
American Museum of
Natural History

A reader's reply to this story provided opportunity for Brown to briefly speculate on the origin of the spider's glow.

Luminous Spiders

The issue of August 21 contains a very interesting letter from Barnum Brown on his discovery in Central Burma of a luminous spider whose abdomen glowed with light while "fireflies sparkled here and there." May not this be analogous to the effect obtained by that prank of childhood when we caught and fed fireflies to the ordinary hop toad and then turned him loose on the lawn in front of the veranda to the consternation of the older folks, who could see but not comprehend the bouncing light. Maybe the spider had feasted plentifully on the abundant fireflies.

Edward Pierce Hulse

In my opinion, there are three explanations of this luminosity: First, the eating of the luminous portions of fireflies by the spiders; second, injection by bacteria or fungi; third, a true luminous organ. Mr. Hulse touches upon the first possibility in his letter and the answer to this, I think, is to be found in the habits of spiders. Spiders are

provided with sucking mouth parts, and do not
devour the material as a whole. If this individual had
selected only the luminous portions of fireflies,
light would have shown not only on the abdomen,
but through the thorax and head as well. I think
the answer to the second possibility, injection by
luminous bacteria and fungi, would hold equally
true, that both thorax and head would have shown
luminosity had the spider been injected. As a matter of fact, I was sufficiently close to determine
accurately that only the abdomen glowed. I think
that it was definitely provided with a true luminous organ.

Barnum Brown

Brown correctly notes the difficulty with explaining away the
spider's luminous character as a result of predation on fireflies.
The possibility of disease (infection with luminous bacteria)
can't be overlooked, but the regular appearance and abdomen-
specific nature of the glow does count against that.
   As a result of this published report, Brown received a letter
regarding another luminous arachnid.

Another Luminous Spider

A note on "A Luminous Spider," published by
me in *Science*, August 21, 1925, it seems has been
copied in the London *Sphere*, and another observation has been reported in a letter from Mr. C.
H. Bompas, Bishops Stratford, Herts, England,
which reads:

I have read your note on a phosphorescent spider from Burma in the *Sphere*.

As you are presumably interested in such things you may like to know that I have seen such a spider at Shillong, in Assam.

The spider is truly phosphorescent and switches on its light when frightened. It is some time since I saw one, but my recollection is that the light came from six or eight spots under the abdomen.

The one I saw was in the middle of a bush and when approached or shaken glowed more brightly, no doubt as a means of defense.

The one I saw was in the middle of a bush and when approached or shaken glowed more brightly, no doubt as a means of defense.

The locality from which this second occurrence is reported is about one hundred miles from the place of my observation in Burma. While the observation differs in many respects, it is, I think, well worth recording.

Barnum Brown
American Museum of
Natural History,
New York, N.Y.

Shillong is the capital city of the northeast India state of Meghalaya. Both of these regions (in Burma and India) are heavily forested, providing ample room for unique invertebrates to remain hidden from the prying eyes of Western observers. The question, though, is really not whether a glowing spider could remain undescribed and uncatalogued, but whether such an animal could exist at all.

While it may seem improbable at first, there are precedents for phylogenetically disjunct bioluminescence in terrestrial invertebrates. Only one bioluminescent land snail is known, the

Malaysian species *Quantula* (*Dyakia*) *striata*. The Brazilian cockroach *Lucihormetica fenestrata* is the only known bioluminescent orthopteroid insect. A small group of closely-related species in the California genus *Motyxia* (*Luminodesmus*) are the only recognized luminous millipedes.

Just as interesting are the luminous "caterpillars" currently being studied by the Springbrook Research Centre, Queensland, Australia. These "caterpillars" (this is, apparently, the best guess as to what the critters are) have glowing blue-green tails, and this luminous glow expands to include the head and digestive tract upon disturbance. Luminous pupae have also been found, but an adult form has not yet been seen. What is intriguing is that early settlers in the area wrote about glowing moths in this sub-tropical rainforest.

But what purpose might bioluminescence have for a spider? Don't spiders require cryptic appearances so as not to warn away prey? Actually, arachnologists know that most kinds of spiders do not merely sit and wait for random hapless insects to come along as dinner. Many species actively lure their prey. In some cases, as in various bolas spiders, chemical cues that mimic lepidopteran sex pheromones attract moths. Several spiders (certain pirate spiders and salticids) lure other spiders by plucking webs to mimic entangled prey.

Besides "scent" and vibratory lures, visual lures are also found in arachnids. The coloration and patterning of certain orb-weavers, within a pollinating insect's range of the visual spectrum, look like flowers. Even the spider's web itself can lure insect prey, as it reflects UV light in a forest setting, creating the illusion of a patch of sunlight. It isn't difficult to imagine luminescence as a successful lure in the forested night, just as it is used by predatory fishes angling in the depths of the ocean.

Unfortunately, until a specimen is located, the Case of the Glowing Spider remains murky. This is one of the few cryptids that might actually be caught by hand—if you can land a grant to cover traveling expenses...

References:

Brown, Barnum. A Luminous Spider. *Science*,
    LXII(1599), Aug. 21, 1925, p. 182.
Brown, Barnum. reply to Luminous Spiders. *Science*,
    LXII (1606), Oct. 9, 1925, p. 329
Brown, Barnum. Another Luminous Spider. *Science*,
    LXIII (1632), April 9, 1926, p. 383.
Hulse, E. P. Luminous Spiders. *Science*, LXII (1606),
    Oct. 9, 1925, p. 329
Maguire, G. 2004. http://maguires.com/research/
    luminous_caterpillars.htm

The Suiformes:
The Expanding World
of Pigs and Peccaries

Matthew A. Bille

In 1905, American humorist Ellis Parker Butler published a short story entitled "Pigs is Pigs." It concerned a railway bureaucrat who insisted a guinea pig was just another pig, "nationality" not withstanding. While the inhabitants of the suborder Suiformes do share some common points of appearance, recent decades have taught us there are many more species and subspecies in this group than once suspected.

The Suiformes include three families, all of which are of some interest to cryptozoology. The hippos (family Hippopotamidae) have held steady at two species for over a hundred years, but there is still some mystery about them. The family Dicotylidae includes the peccaries, and they are of somewhat more interest, for reasons we will come to shortly.

Finally, there are the "true pigs," or suids. The family Suidae includes the genus *Babyrousa* (one species), *Hylochoerus* (one species, *Hylochoerus meinertzhageni*, the giant forest hog, which weighs up to 600 lbs), *Phacochoerus* (two species), *Potamochoerus* (two species), and *Sus* (ten species, including the omnipresent *Sus scrofa*, to which all domesticated pigs and the ancestral wild boar belong). This includes some distinct surviving subspecies like *S. s. riukiuanus* of the Ryukyu Islands.

The true pigs are Old World creatures. Their counterparts in the New World are the peccaries. Peccaries, now restricted basically to Central and South America (one species ranges into the southwestern United States), did live in Eurasia millions of years ago, and they were common in North America until the late Pleistocene. One of the features pigs and peccaries have in common is unique among mammals. This is the rhinarium, or snout disk. The snout has a circular tip, reinforced with cartilage, used for rooting in the ground. The hippos, the only genus in this subfamily lacking the rhinarium, are also Old World types, confined to Africa and Madagascar. The traits differentiating peccaries from pigs include differently shaped upper canines, shorter tails, the structure of the hind foot, and a scent gland peccaries have on their backs. (Oddly, the tooth structure of peccaries and hippos shows many similarities, leaving intriguing questions about their relationship.)

The Suiformes have served as human food and livestock for millennia. In the Niah Caves of Sarawak, human remains surrounded by the bones of bearded pigs have been dated to 40,000 years BP. From Europe to Indonesia, wild and domesticated pigs have formed vital links in the human food chain. South American natives hunted the peccaries as well. Hippos have likewise been exploited for food and ivory, but their localized range and sheer size have prevented them from becoming a staple food source or a domestic animal.

New pigs and peccaries enter the textbooks in three ways: by reclassification of existing specimens; by rediscoveries of species thought dubious or extinct; and, most excitingly, when entirely new species are found in the wild.

The interesting member of the hippopotami is *Hexaprotodon liberiensis*, the pygmy hippopotamus. This solitary, nocturnal animal, which may weigh up to 600 lbs, was regarded as a myth by Western science until the mid-nineteenth century. It hid out in the dense, swampy forests and rivers of Western Africa until described in 1849, and remained in an uncertain status long after that. Not all zoologists accepted it as a valid species until live animals were returned to Europe in 1911.

Could any hippos still be overlooked? A dwarf form on Madagascar (*Hippopotamus lemerlei*) is believed to have been hunted to extinction (as were two other Madagascar hippo species). This supposedly took place 800-1,000 years ago, but paleoecologist David Burney was told by villagers in western part of the island that an animal of similar description, locally called *kilopilopitsofy* (meaning "floppy ears"), was still seen, or at least had been seen in their lifetimes. In a 1999 paper, Burney wrote that accounts described, with great consistency, an animal which "is nocturnal, grunts noisily, and flees to water when disturbed... it is cow-sized, hornless, dark in color, and has a large mouth with big teeth."

In a 2004 paper on dating fossil and subfossil remains from the island, Burney added the important note that, "One specimen of *Hippopotamus* of unknown provenance dates to the period of European colonization." A surviving *kilopilopitsofy* has yet to be located: however, the island's forests, even though endangered by over-cutting and development, still include large, rugged areas for which there are no maps (and no roads to show on maps if the latter existed.) Cryptozoologists hold out hope that at least one member of the island's prehistoric megafauna still lives.

The known peccaries, for a long time, consisted of two species. It was quite a shock to zoologists when a third species, the Chacoan peccary (*Catagonus wagneri*), turned up alive. Skulls of recently deceased animals were collected by a mammal survey team led by Ralph M. Wetzel in Peru's remote Grand Chaco region in 1972. A formal paper followed in 1975. The animal had been described from subfossil remains in 1930, but no one had suggested that herds of live 70- to 90-pound animals were being overlooked.

Dutch primatologist Marc van Roosmalen, known for discoveries of several new monkey species in the Amazon basin, believed there was an unclassified type of peccary at large there as well. In 2000, he wrote that he had seen a new peccary twice. This peccary lived in small family groups and sometimes joined herds of the two known species of this region, the collared peccary or javelina (*Tayassu tajacu*) and the white-lipped peccary (*Tayassu pecari*). Van Roosmalen's peccary appeared to be intermediate in size between the other two species, which put the adult weight in the range of 40 to 60 pounds.

Van Roosmalen described the new animal as "dusty brown." The white-lipped peccary may be brownish, but has distinctive white markings. The collared peccary is gray to black.

In 2003, Van Roosmalen's Web site, AmazonNew Species.com, included the entry: "New species of peccary: Temporarily removed until publication in print journal."

In 2004, Van Roosmalen and German filmmaker Lothar Frenz happened to be present when a strange-looking peccary was turned into spare ribs by villagers in the Rio Aripuanã area. The visitors did not partake, but saved a portion for DNA testing. That sample, combined with the subsequent filming of a live specimen, was enough for Van Roosmalen to propose the new species *Tayassu maximus*, the giant forest peccary.

Than animal is larger than van Roosmalen first thought, weighing up to 90 pounds. That makes it similar in size to the Chacoan peccary, which confusingly has been referred to as the "giant peccary" as well.

At this writing, the formal description of this new species has yet to be published. Dr. William Oliver, chair of the Pig, Peccary, and Hippo Specialist Group (PPHSG) of the International Union for the Conservation of Nature (IUCN), cautions that "is a long way from being confirmed as a new species." Nonetheless, the PPSHG thought the discovery potentially significant enough to place an announcement and photograph on its Web site.

Suid taxonomy is a bit confusing because of widespread domestication. Domestic pigs come in countless varieties, and all will mate with their wild relatives. Feral *Sus scrofa* and those of mixed wild and domestic ancestry are common in Eurasia and Africa, where wild suids originated, and in the Americas and Australia, where all are from imported stock. Some of these attain impressive sizes, as in the case of "Hogzilla," the monster killed in Georgia in 2004. Alleged to weigh 1,000 lbs, this storied beast attracted an all-out scientific investigation backed by *National Geographic*. Researchers who excavated the remains determined the animal, a hybrid wild boar-Hampshire pig, probably weighed around 800 pounds – still a lot of bacon.

These animals can be ruinously destructive to local ecosystems. In Hawaii, they are blamed for making some rare plants and birds endangered or extinct. On the island of Santa Cruz, off California, they are being hunted down by wildlife officials attempting to save the endangered Santa Cruz fox.

Even pigs known to be of wild ancestry may look very similar, and the list of approximately ten full species is periodically subjected to modifications.

A notable rediscovery of a species in limbo took place a decade ago in Southeast Asia. The Vietnamese warty pig, *Sus bucculentis*, had been described in 1892. At that time, Father Pierre-Marie Hende, a Jesuit missionary, literally drew the attention of science to the animal by publishing a sketch of its skull. While two skulls of the warty pig were eventually obtained from southern Vietnam, no live specimen was ever seen by a Western scientist.

The species seemingly disappeared in the wild, and was long considered extinct or even invalid. Dr. William Oliver, head of the IUCN/SSC Pigs and Peccaries Specialist Group, wrote in 1992 that, "The species is probably endangered, if not already extinct."

On a 1995 Wildlife Conservation Society expedition into the forested hills of Vietnam and Laos (which produced a new deer, the giant muntjac, along with the rediscovery of Roosevelt's muntjac), zoologist George Schaller and Laotian scientist Khamkhoun Khounboline were told of a yellowish-furred pig with a long snout. The two men succeeded in obtaining the partial skull of a freshly killed juvenile. While the 1996 IUCN Red List continued to classify *S. bucculentis* as Extinct, a search was underway for evidence that the new skull might be proof of its survival. After locating one of the long-missing type skulls in Beijing in 1996, George Amato and Colin Groves succeeded in matching the recent evidence to the description of *S. bucculentis*.

The species may be extinct in Vietnam, since the rediscovery took place in Laos, farther north along the rugged Annamite mountain range. It certainly is not very numerous, and conservation efforts are underway.

Another relatively recent tale of rediscovery concerns the pygmy hog (*Sus salvanius*), described in 1847. Originally a resident of northern India and Nepal (its specific name refers to Nepal's Saul Forest), the animal was last known from Assam in India. By the end of the 1950s, though, sighting reports had almost ceased. According to the IUCN, its status in 1965 was, "Very rare and believed to be decreasing in numbers." Some authorities wrote the animal off entirely. The dark-brown pig, which weighs only 15 to 25 pounds as an adult, was either endangered or possibly extinct.

The species was rediscovered in 1971 by tea planters in northwestern Assam after the burning of grasslands drove some individuals from their hideaways. At least nine live specimens were captured (a few more were no doubt captured but cooked rather than being presented to science).

Extinction-wise, the pig is still in a precarious state. The IUCN classifies the animal as Critically Endangered (despite a captive breeding program which has had some success), and the current wild population many be only 150 or so. Still, even being in great danger is a vital improvement over being extinct.

Then there are the reclassifications of suids which were right under our noses, but were inaccurately lumped together.

In a 1998 article in the journal *Species*, William Oliver reported on a reappraisal of the genus by anthropologist and taxonomist Colin Groves. Dr. Groves reported, to begin with, that there were more wild pigs in the Philippines than previously believed. He found there were two previously-undescribed species, the Visayan warty pig (*Sus cebifrons*) and Philippine warty pig (*Sus philippensis*). While Groves made his determination based on morphology, subsequent analysis has shown these species have 36 chromosomes, as compared to 38 for the species both were thought to belong to, *S. celebensis*.

An interesting note on the Visayan pig: until it was bred in captivity as part of a conservation program, it was so totally unknown that even its external appearance was uncertain. Now we know the adult males have prominent manes and crests to make themselves look more imposing, combined with an array of warts on their jaws and cheeks.

Today, *S. cebifrons* is critically endangered, its habitat reduced to a few islands. Some specimens have been shipped overseas. The St. Louis zoo is one of the institutions harboring these pigs for breeding as part of the of an American Zoo and Aquarium Association (AZA)-sponsored Species Survival Plan (SSP).

In 2005, it was announced another Filipino subspecies, the Palawan pig (*Sus barbatus ahoenobarbus*) should be elevated to a full species. It appears to be more closely related to other pig species of the Philippines than to its previously assumed species, the bearded pig *Sus barbatus*. With three wild suid species endemic to the Philippines, this island group may be tied for the title of "hog heaven" with Indonesia, which also has at least three endemic species.

Further study may reveal more species yet. Groves wrote in September 2005 that he wondered, "For example, are the Visayas warty pigs really all one species? The Palawan bearded pig was recently upgraded to species status, and the pigs of Jolo, in the Sulu archipelago (southern Philippines), have never been properly classified and I'm told that they are probably a distinct species. I have for a long time been uneasy about my own lumping of all the Eurasian pigs into one species, *Sus scrofa*."

Writing for the IUCN PPHSG in 2002, researcher Desmond Allen reported that pigs on Tawi Tawi and other southern Philippine islands were still as unknown as they were endangered. (The danger to pigs on Tawi Tawi comes mainly from deforestation: this particular suid population is fortunate in that the local humans are Muslims and do not hunt pigs for food.) Allen mentioned that a colleague, Karen Rose, had visited the area in 1997 and reported seeing one pig that was "probably an unknown species." The species of pigs on the island of Tablas (where they are being hunted) is still unknown.

Thanks to extensive research by Peter Grubb, African pigs have also undergone a reevaluation. Grubb confirmed the suggestions of some earlier experts that the red river hog (*Potamochoerus porcus*), which is sometimes hunted by humans despite its charming habit of grubbing tree seeds out of elephant dung, and the bushpig (*Potamochoerus larvatus*), were clearly separate species, although they had commonly been described as one.

Groves wrote in 2005 that, "When Peter Grubb looked at the skulls of warthogs, he deduced that there were two species, and an extinct South African species still existed in the Horn of Africa – which in my book makes it rate as a real new species." This last was the Cape, or desert warthog (*Phacochoerus aethiopicus*), believed extinct for over a century. In the course of Grubb's long labors in specimen study, surveys of local wildlife officials and scientists, and fieldwork, he realized warthogs in northern Kenya in in Somalia were distinct from the common warthog, *P. africanus*. They were not, he determined, an unknown species, but the long-written-off *P. aethiopicus*. The "new" species has, among other characteristics, a different number of incisors than the common warthog. (It is often true that teeth offer major clues to mammalian classification).

In a 2002 paper, other authors using DNA analysis confirmed and further clarified that the desert warthog was a "genetically deeply divergent species." Indeed, while it and the common warthog had been confused visually, the two had been distinct since the Pliocene.

What might still be out there? Groves writes, "There are constant reports of Giant Forest Hogs in Ethiopia: are they really there, and are they really GFHs? Then look at the unknown hotspots in Asia: we now know about the Annamites, but another, just beginning to yield its secrets, is Arunachal Pradesh."

It has been suggested the bizarre babirusa (*Babyrousa babyrussa*) may also be more than one species. (Some researchers place this unique animal in its own subfamily, Babyrusinae, and it has even been suggested it merits its own family.) The babirusa is a strong contender for the title of "ugliest mammal alive." It can weigh over 200 lbs and is almost hairless, with large folds of skin on the neck and belly. The upper canines of the males may be a foot long. They grow up through the snout, piercing the skin, and curve backwards. Added to the lower canines, which erupt from the sides of the mouth, this dental equipment is like nothing else in the animal world. Despite their fierce appearance, the animals can be tamed if captured when young.

This suid is known from Sulawesi and nearby islands, with three subspecies recognized as living and one (*B. b. bolabatuensis*) as recently extinct. Groves, writing in 2001, proposed upgrading all four to species level, which would create as species the diminutive *B. bolabatuensis*, the sparsely haired *B. celebensis*, and *B. togeanensis*, a large race confined to the Togean Islands, in addition to *B. babyrussa* with its distinctive long, thick coat. The type species (or subspecies, *B. b. babyrussa*), was missing and feared extinct for three decades before turning up in the late 1980s. Babirusas used to exist all over the large island of Sulawesi, but are now confined to scattered areas in the north, central, and southeastern areas. It is still unclear whether any undiscovered and/or unclassified populations still hold out on remote islands. The Indonesian archipelago includes many islands never visited by scientists, and remote populations could have been diverging for thousands of years.

The Suiformes deserve better than the disdain shown by many human populations for anything labeled a "pig." They are an ancient, worldwide race of mammals, often serving as a key human food source in addition to the roles they play in their own ecosystems. They have been domesticated, not only for food, but for such exotic occupations as hunting truffles. The Suiformes include species we have only recently met, species which have been or yet may be rediscovered, and, almost certainly, some species humans have never met. Pigs is not, after all, just pigs.

References:

Allen, Desmond. "Some poorly-known pigs in the Philippines: Tawi Tawi and Tablas," *Asian Wild Pig News* (IUCN PPHSG newsletter, now renamed *Suiform Soundings*), Vol. 2, No.1 (January 2002).

Burney, David, and Ramilisonia; "The Kilopilopitsofy, Kidoky, and Bokyboky: Accounts of Strange Animals from Belo-sur-mer, Madagascar, and the Megafaunal 'Extinction Window'," *American Anthropologist*, 100:957 (1999).

Burney, David, et. al. "A chronology for late prehistoric Madagascar," *Journal of Human Evolution* 47:25-63 (2004).

Groves, Colin. Email to Matt Bille, September 12, 2005.

Groves, Colin P. "Taxonomy of wild pigs (Sus) of the Philippines," *Zoological Journal of the Linnaean Society*, 120:163-191 (1997).

Groves, Colin P., et al. "Rediscovery of the wild pig *Sus bucculentis*," *Nature* (386), p.335 (1997).

IUCN, Pigs, Peccaries and Hippos Status Survey and Action Plan (1993). See the PPHSG home page at http://www.iucn.org/themes/ssc/sgs/pphsg/ for updates. For the group's newsletter, *Suiform Soundings*, see http://iucn.org/themes/ssc/sgs/pphsg/Suiform%20soundings/Newsletter.htm.

IUCN Red List, 2004. Available at www.redlist.org.

Lucchini, V., E. Meijaard, et al. "New phylogenetic perspectives among South-East Asian wild pig species based on mtDNA sequences and morphometric data." *Journal of Zoology* 266(1): 25-35.) (2005).

Macdonald, Alastair. "The Babirusa (*Babyrousa babyrussa*)," IUCN Pigs, Peccaries and Hippos Status Survey and Action Plan, Chapter 5.8 (1993).

Meijaard, Erik. Email to Matt Bille, September 13, 2005.

Meijaard, Erik, and Colin Colin Groves, "Proposal for taxonomic changes within the genus *Babyrousa*," *Asian Wild Pig News*, Vol. 2, No.1 (January 2002).

Oliver, William. Email to Matt Bille, September 12, 2005.

Oliver, William. "More and More and Fewer and Fewer," *Species*, No. 30, June, pp.57-8 (1998).

Randi, Ettore, et. al. "Evidence of two genetically deeply divergent species of warthog, *Phacochoerus africanus* and *P. aethiopicus* (Artiodactyla: Suiformes) in East Africa," *Mammalian Biology*, Volume 67, Number 2, pp. 91-96(6) (2002).

Tyson, Peter. *The Eighth Continent: Life, Death, and Rediscovery in the Lost World of Madagascar.* New York: William Morrow (2000).

Van Roosmalen, Marc. Personal communications to Matt Bille, January 4 and March 2 (2000).

Van Roosmalen, Marc. Descriptions on Web site, "New Species from Amazonia," http://www.amazonnewspecies.com/index.htm (2003). (Note: Site was down as of October 2005.)

Coelacanths of Silver: A Biogeography of Latimeria?

Gary Mangiacopra
Dwight G. Smith

Among those investigators who pursue cryptozoology, it has become a cliché to cite the case of the coelacanth, a crossopterygian (lobe-finned) fish which was supposed to have become extinct some 65 million years ago, yet was rediscovered alive and apparently quite well in the third decade of the 20th century. The discovery of this living fossil prompts both implications and stimulus for the field of cryptozoology, not least because of the circumstances attendant to its discovery.

The story begins with 32-year old Marjorie Latimer, Curator of the Museum of Natural History located in the port town of East London, South Africa, and her friend, Captain Hendrick Goosen of the sea trawler Nerine. The captain often spent weeks fishing the coastal waters of this corner of the Indian Ocean and frequently brought in unusual fish specimens he had captured for Miss Latimer. On 23 December 1938, Captain Goosen brought his trawler into port with a load of fish caught off the mouth of the Chalumna River. Amongst the fish piled on a dockside table, Miss Latimer spotted a large blue protruding from the bottom. She pulled the enormous fish—nearly five feet in length—out from beneath the pile and, in her words saw "the most beautiful fish I had ever seen, five feet long, and a pale mauve blue with iridescent silver markings."

Marjorie had no idea what it was, but was sufficiently in-trigued to rescue the monster fish for her museum. She was unable to identify the fish with her reference book collection at the museum, so Miss Latimer sent a photograph and sketches of it to J.L. B. Smith, a chemistry professor at Rhodes Univer-sity in Grahamstown. Chemistry professor notwithstanding, Smith immediately recognized the fish as a coelacanth and cabled Miss Latimer to hold on to the specimen: "Most impor-tant preserve skeleton and gills = fish described."

The rest, as the cliché goes, is the stuff of history. The dis-covery of an animal presumed extinct for at least 65 million years caught the public's fancy. A photograph display of the specimen draw 20,000 visitors on the first day of its showing in London, England. The coelacanth discovery stimulated the search for other specimens; pictures were posted in many of the fishing villages of southern Africa and a reward was offered for another specimen. Despite the powerful inducement, it took another 14 years before the second coelacanth specimen was discovered, this time caught by a local fisherman using a hand-line in the waters just off the Comoros Islands, a group of islands not far off the northwest coast of Madagascar.

Since that time more coelacanths have shown up along the southern coast of Africa and—amazingly enough, even more re-cently in reef waters off the coast of North Sulwesi, Indonesia, several thousand miles to the east of the original finds off East Africa. First discovered in 1998, the Indonesian coelacanths are just as big as their Comoros cousins, but are brownish rather than blue in color, with gold flecks of color along their sides. DNA analysis has confirmed that the two coelacanths are dif-ferent species.

A Coelacanth Resume

The two living species of coelacanths are the last surviving species of a once numerous and widespread family of fleshy-finned fishes called the Sarcopterygians. Their ancestors date

Coelacanth

from the Devonian period of 410 million years ago. Fossil coelacanths have been found in geological deposits of all continents
except Antarctica.

Two species of coelacanths known today are considered prehistoric relics. They are named for their hollow fins (Greek:
coelacanth = hollow spine) and also have unique fluid-filled
cartilaginous tubes for a backbone. The two species occur in two
widely separate geographic areas, off the southeastern coast of
Africa and in the Indonesian Islands of the East Indies.

Ecologically, the coelacanths are reef dwelling predators of
comparatively large size. A female coelacanth taken off
Mozambique is well over two meters in length (well over six feet)
and nearly 100 kilograms in weight (more than 150 pounds).
Although their ecology is still poorly known, coelacanths are
apparently opportunistic feeders or scavengers on a wide variety of reef invertebrates such as squid and reef dwelling fishes
including lantern fish, witch eels, cardinal fishes, and snappers.

The relationship between coelacanths and the first tentative
amphibians to "crawl or clamber" ashore onto land is problematic, although some paleontologists suggest that the lungfishes
were the immediate ancestors of the first amphibians. Whether
modern coelacanths use their "pectoral and pelvic fins to walk

along the reef" has not yet been observed. Coelacanths are also distinguished as live bearers—that is, the eggs hatch within the female and the young emerge alive from her brood pouch.

Population and distribution of modern coelacanths is also poorly known. Ecologists suggest a very rough population estimate of about 1000 individuals in two populations, one centered along the southern African coast from Mozambique southward to East London and the other population located along the coast of North Sulwesi, Indonesia. Concerns that the coelacanth population may be limited to this specific region of the world's ocean coupled with request for specimens (promoted by a bounty for them) and the quest to obtain a living specimen for an aquarium exhibit, may push this species to extinction in the age of humans (Brown, 1988). Pending further information, the coelacanth is listed as an endangered species in Appendix I of the CITES listings.

The Coelacanth-Crytozoologist Connection

The fabulous discovery of living coelacanths has proven to be a bellweather phenomena for cryptozoologists. Many—very many—cryptozoologists and other wildlife investigators repeatedly make the case that if this long assumed to be extinct species of fish turned up alive then perhaps other stranger and equally "assumed extinct" animals (or plants) might yet be lurking within the depths of the oceans or roaming within the deep, dense and little explored forests of the world. Cryptozoologists also point out that coelacanths were apparently well known to natives long before they were discovered by scientists. They were the *M'tsamboidoi* of the Kombessa to native fisherman of southern African and Madagascar. Their flesh was eaten and their scales were even used to roughen the edges of bicycle tires prior to patching them. The Indonesian coelacanths were called *Rajah Laut*, or "King of the Sea" by the Sulwesian fishermen that occasionally snared one in their nets or hauled them in on deep drag lines. Thus oddly, it was not how long this "living fossil" evaded man—but rather how civilized whites overlooked it!

Cryptozoologists attach considerable importance to the connection between folklore and discovery. If tales of coelacanths can turn out to be true then why can't we give equal credence to local legends of living dinosaurs, yeti, and other fabulous creatures of native folklore? Not, of course, as scientifically established creatures but rather as potential investigative entities of the type that form the foundation for the science of cryptozoology.

The Biogeography of Coelacanths

In the last 30 years or so, a number of small silver Spanish models of what appear to be coelacanths have turned up. Their discovery has generated a controversy of sorts concerning both the range over which coelacanths occur in the wild and the length of time that coelacanths have been known. Is it possible, for example, that these coelacanths in silver actually depict a Mediterranean population that did exist or still exists? In other words, was the Mediterranean the source used by ancient silversmiths to craft their likeness of silver coelacanths? A similar controversy concerns the existence, past or present, of a population of coelacanths in the Gulf of Mexico.

Only a decade prior, neither of these coelacanth geographic scenarios were thought to be likely, but the discovery of a second population of coelacanths in Indonesia, several thousand miles (nearly 10,000 kilometers) removed from the Comoros Island populations of southern Africa has considerably clouded the issue of exactly where coelacanth populations occur. Could occasional strays wander around southern South Africa and into the South Atlantic and finally reach northward as far as the Caribbean Sea or the Mediterranean? Professor J. L. B. Smith, the world's expert on this fossil fish investigated this possibility for many years before his death in an effort to establish a modern biogeography of a very ancient species.

Any biogeography of coelacanths must begin with the place of their original discovery, off the Comoros Islands. These islands lie directly east of southern Africa and slightly northwest of Madagascar. After reporting the first specimen of "surviving fossil" in the pages of *Nature* and naming it *Latimeria chalumnae*, J. L. B. Smith began receiving additional reports of coelacanths from along the South African coast (Smith, 1939, 1940):

> "A reliable citizen-angler of East London has stated that he had some five years previously come across exactly such a fish, only considerably larger, partly decomposed, cast up by the waves on the shore some miles east of East London. He described the specimen in such terms as to render it very likely that he was not mistaken, and stated that when he returned with assistance to fetch the monster, it had disappeared with a risen tide. Also, though probably less reliable than the above, is a statement emanating from some member of the crew of a fishing vessel, that one of them had seen no less that six such fishes taken in on one haul near Durban. All had been discarded along with other unwanted fish in the trawl."

This would place the East London citizen-angler beach discovery circa 1934 or 1935; and the trawl-captured specimens in the early 1930s. To secure additional specimens, Professor Smith offered a reward of 100 pounds sterling. Almost immediately he received a verbal report of a capture from the Mozambique Channel (Smith, 1956):

> "In 1948 I met a native in the Bazarutol area of Mozambique who picked on the fish at once... In the water it was like a big Garrupa (rock cod), but when he got it out the big scales and the peculiar fins stamped it on his memory as unique. He spoke of its oiliness, the soft flesh, and the absence of bone, very unfish-like characters about which he never could have known except from an actual

specimen. He could not say if the tail was the
same, but it was near enough."

Two decades later, in 1965, Smith received correspondence
from a Mr. Bert Spring, a lighthouse keeper at the Cape St. Blaize
near Mossel Bay who claimed to have seen a coelacanth swimming near the water surface out in the bay around the lighthouse (Bruton, 1989).

An even more remarkable coelacanth discovery began with
a tale related by coelacanth Researcher Michael Bruton (1989):

> "Recently I had the privilege to visit the old
> boardroom of the famous shipping and timber
> company in Knysna, courtesy of the well-known
> novelist, Hjalmar Thesen. He showed me a paint-
> ing by his uncle, Leonard Thesen, which quite
> clearly depicts a coelacanth, resplendent in its
> blue with white flecks, along with other common
> fishes from that area such as the leevis, kob and
> parrot fish. The painting was originally done on a
> piece of curtaining for a cottage purchased in 1924
> but has since been framed and covered with glass.
> A label on the glass states that the painting dates
> from 1925, which would be extraordinary if true."

Leonard Thesen died in 1951, leaving unanswered the circumstances surrounding the astonishing painting. Where and
when, for example, did Thesen see the coelacanth? His relatives
were unable to provide answers. They knew that Thesen was
"an enthusiastic beachcomber" so perhaps he was lucky enough
to come across a dead specimen or specimens that had washed
up on shore?

Years later, in 1994, Thesen's nephew, Hjalmar Thesen
added a few more details about the history of the painting which
had since been given to a local fishing museum in Knysna.
Hjalmar candidly informs us that Thesen's sister remembered
Leonard being irritated by all the fuss about the sensational

coelacanth discovery of 1938, saying that "he has seen one of those already!"

Professor Smith continued to receive reports of coelacanths from writers, fishermen, and wildlife enthusiasts who claimed to have seen or caught coelacanths in other parts of the world. An American soldier stationed in the Korean Peninsula confidently stated that coelacanths were commonly sold in South Korean fish-markets. Another woman reported that one was offered to her by a Bermudian fisherman. Smith, however, took the position that there were no coelacanths in the North Atlantic (Smith 1956).

Less than two months before Smith acquired a second coelacanth specimen, a Salisbury resident, Mr. Geoffrey F. Cartwright, wrote a lengthy letter describing an underwater encounter he had with a living coelacanth in the waters off Kenya (Smith, 1953):

> "In October and November of last year (1952) I spent a holiday at Malindi in Kenya. You were conducting research work at Causarian Point and I met your wife on the beach one day as I and several others returned from hunting under water... I occupied myself a great deal in goggle fishing and hunting under water and consider that I saw most of the fish in the area, for I visited a great number of coral reefs in the area. Why I mention all of this is to indicate that I have gained a fair knowledge of various types of fish and I found that most of them did not appear to carry scales but were of smooth appearance, or the scales were light. One day I had swum out with my harpoon gun. I looked down into the water and just below my sand-shoed foot, I saw a large fish. It was heavily built and probably weighed from 100-150-lbs. I thought how just too comfortably my foot would fit into its mouth. It was totally unlike any other fish I had seen or saw afterwards. It looked wholly evil and

a thousand years old. It had a large eye, and the most distinguishing feature was the armour-plated effect of its heavy scales—scales so heavy that it was set quite apart from all other fish I saw. It had a baleful and ancient appearance, dull, dark and gray. I decided that I would attempt a head shot and with luck, might secure the fish. My gun was not quite as powerful as it might have been as it had only two rubbers. Nevertheless, I had caught quite a considerable number of fish. On this occasion the harpoon struck the fish but did not penetrate. The fish cleared."

Cartwright made several inquires among Malindi residents as to what he had seen, and accepted the suggestion that he had seen a large rock cod—though one that was a different species that the type he had found among the corral reefs which were small bottom dwellers (6-lbs) black, brown or yellow with brown speckles whose mouths were as wide as their bodies.

At first Cartwright disregarded the possibility that he had seen a living coelacanth—after all, he had seen a plaster cast of a coelacanth at the Nairobi Coryndon Museum in 1950 and did not seriously consider a possible resemblance. However, during the Christmas season of 1952 a second coelacanth specimen was acquired by Smith and Captain Hunt (the fishermen captured it at a depth of about 120 feet only 600 feet from the shore (Thomson, 1991). Upon seeing the photographs of the second capture Cartwright was struck by the likeness to the fish he had observed. A year later, Cartwright visited the Rhodes Centenary Exhibition in Bulawago and examined the plaster coelacanth model exhibit, taking care to stand in front of the fish, looking downwards from the front into it's eyes, which was exactly the same position from which he had observed the big fish at Malindi. After a lengthy inspection, Cartwright was convinced that he had seen a coelacanth after all.

Professor Smith enthusiastically endorsed Cartwright's suggestion that he had actually spotted a coelacanth. In a lengthy reply to Cartwright, Professor Smith pointed out that very few East African fish actually reached 100 pounds in size, and none bore the least similarity to the heavy-bodied coelacanth. Smith concluded, "So your fish, having been a grayish color is not incompatible with it's having been a coelacanth."

The 1952 capture of the second coelacanth specimen resulted from the efforts of Captain Erik Hunt. Following up rumors of big fish being caught off the East African coast, Captain Hunt posted reward notices in the Comoros Islands region, northwest of Madagascar. Native fishermen sometimes took a large fish they called the "Gombessa" and both Professor Smith and Hunt strongly suspected that Gombessa was actually a coelacanth. The 1952 capture was followed by a number of subsequent coelacanth catches, all from the general region of the Comoros Islands.

The concentration of coelacanth captures and observations ultimately led Smith to suspect that the Comoros Islands were the distribution center of the coelacanth population and that the first—and later—coelacanths captured along the East African coast had drifted southward with the currents through the Mozambique Channel.

Coelacanth records have continued to accumulate. On August 11, 1991, a coelacanth was caught in a trawl off Maputo, Mozambique on August 11, 1991. Between 1991-2001 at least four and possibly more coelacanths were netted off the Madagascar coast. Captures of these coelacanths raised further questions about the biogeography of the species. Were these simply "drifters" or did they represent population centers distinct from the Comoros Island coelacanth population?.

An even more astonishing find took place in 1997—46 years after the first coelacanth discovery—when Mark and Arnaz Erdmann spotted a coelacanth which had been netted by fishermen off Manado Tua Island, North Sulwesi, Indonesia. They took photos and placed them on the Internet. Supported by a National Geographic Society and Smithsonian grant, the two returned to the Indonesian island the following year to resume their search for more coelacanths. They were rewarded on July

30th, 1998, with the capture of a second specimen, also taken from the same general area. DNA samples revealed that these Indonesian coelacanths were sufficiently distinct from the South African coelacanths to warrant a new species designation, *Latimeria menadoensis.*

Still more interesting revelations were on the way. On 28 October 2000, scuba divers Pieter Venter, Peter Timm, and Etienne le Roux discovered a coelacanth in 104 meters of water (about 320 feet), thereby disproving the long held tenet that coelacanths were always deep reef fishes. The discovery took place in the waters of Sodwana Bay, which is part of the St. Lucia Marine Protected Area, geographically located just south of Mozambique. On a second dive, a five man team discovered three coelacanths as they explored the caverns along the reef. They estimated the largest at 1.8 meters in length and noted that the other two were smaller, perhaps 1.2 meters and 1.0 meters in length. The coelacanths took no notice of them while swimming head down, apparently feeding along the ledge. In March and April of 2002, another team effort in the same area using the Jago Submersible was rewarded by the observations of at least 15 coelacanths, one decidedly pregnant.

The startling discovery also broadened the known distribution of coelacanths along the South African coast and raised still more questions about the biogeography of this species.

A Coelacanth Biogeography in Silver?

While vacationing in Spain during the summer of 1964 a chemist, Dr. Ladislao Reti, purchased a 4-inch long silver ornament of a fish that bore a striking resemblance to a coelacanth. Dr. Reti found the silver model hanging in a village church near Bilbao, on the north Atlantic coast of Spain. He believed that the model was made by a Spanish silversmith at least a century prior, which—of course—would be long before the first coelacanth specimen was discovered by Miss Latimer and Professor Smith in South Africa. Donald P. DeSylva offered up to possible explanations for this unique find in the journal *Sea Frontiers*. He thought that perhaps the silversmith used a fossil coelacanth as the model for the silver votive. He also ventured the considerably more exciting suggestion that the silver coelacanth might be based on a coelacanth specimen taken in the nearby waters of the Western Mediterranean or the Eastern Atlantic—specifically the vicinity of the Canary Islands and the Azores. In 1966, Professor Smith offered an alternative opinion to this silver coelacanth (Bell, 1967):

> "If the silver coelacanth is genuine, as it could be, then it could easily have arisen from someone who had voyaged in the early times to the east and seen the fish at the Comoros—for in those days, over a long period, all the ships called at Anjouan Island, one of the larger islands in the Comoros group."

Professor Smith was less enthusiastic about the possibility of an undiscovered Mediterranean or Azore population of coelacanths. Smith noted that the silversmith art had long been important in Africa, especially in coastal ports where traders generally called. Furthermore, the Portuguese were the first European traders in southern Africa and their ships often carried Spanish sailors and Spanish traders as well. So Smith was of the opinion that it was quite possible that a Spanish or Portuguese trader had acquired the silver coelacanth model and taken it to Europe.

Coelacanth enthusiast Bruton (1985) offered his opinion that the mystery of Reti's silver coelacanth may never be solved unless "some angler or commercial fisherman catches a living coelacanth off West Africa or in the Mediterranean."

Renewed interest in a Mediterranean coelacanth population followed the production of a film about the living fossil by coelacanth specialist Hans Fricke. A fish seller who chanced to watch the televised show contacted Frickle and informed him that he had seen a photograph of a coelacanth in a local Portuguese magazine. Fricke also received a phone call from a South American chemist who reported seeing a coelacanth in a fish market of the Balearic Islands (Fricke, 1989; Alexandre, 1994). The Balearic Islands are a small group of islands situated just off the southeast coast of Spain and very much within the realm of the Mediterranean Sea.

The discovery of another silver coelacanth model continued to generate interest in the possibility of a Mediterranean coelacanth population. A university student from Brussels named Maurice Steinert revealed that he had purchased a silver coelacanth model for 9,500 pesetas in an antique shop in Toledo, Spain. Along with the bill of sale, Mr. Steinert was given a certificate attesting that the silver model dated from at least the 19th century (Anthony, 1976). German coelacanth investigator Hans Fricke (1989) examined the antique coelacanth in silver and offered the following comments: "In July 1987... I flew to Brussels: I wanted to see the silvery coelacanth with my own eyes. There I finally had him in my own hands: intricate joints moved the body. His head lets itself flap downward over a joint, and in this way a letter of another thank you gift can disappear into the belly of the fish. But I discover some more. The scales were especially delicately engraved and show patterns which turned out to be spots just like in *Latimeria*." Fins and fin attachments matched those of the famous "living fossil but the head did not greatly resemble that of a coelacanth." And further, "Just like this tail, the typical small extra fin of the *Latimeria* contained only by way of suggestion."

Steinert also showed Fricke several photographs of other silver fish models that had been in the antique shop. He thought that all of the models were from South America and possibly the islands of the North Atlantic. After inspecting the photographs, Fricke made the following suggestions:

> "Some of these churches contain silver votive artifacts in the form of fish. The curious thing about these fish is that they are clearly representative of coelacanths—but not the *Latimeria chalumnae* coelacanths we find off the Comoros. These works of art are carefully detailed and I am

convinced they depict a completely different type of coelacanth, one perhaps smaller than *Latimeria*, that might have lived until recently in the Atlantic or Mediterranean. The question, of course, is whether any might still be alive."

Fricke (1989) originally thought that at least some of the silver models may have been the work of 20th century silversmiths. For example, the pattern of spots on the Steinert silver model clearly resembled the patterns illustrated in the 1954 coelacanth photograph that had been published in a French scientific publication. To Fricke this suggested that the model had to have been produced some time after 1954. Fricke was also quick to note, however, that the silverwork in the Steinert model is far more typical of 17th or 18th century craftsmanship. More recently Fricke and Plante (2001) arranged for the Spanish silver coelacanths to be examined by Professor Cruz-Valdovinos of the University of Madrid and A. Jimenez, a silver expert from Madrid. Both were of the opinion that the famous Spanish silver coelacanths are neither very old nor from Meso-America. On this basis, Fricke and Plante (2001) stated that the "silver coelacanths from Spain are not proof of a pre-1938 discovery and do not provide evidence of a new biogeography of this species outside the Comoros." Of course, the latter conclusion has proven moot with the discovery of an Indonesian coelacanth.

Dr. C. Carpine (1993), curator of the Oceanographic Museum in Monaco, offered the possibility of a third silver coelacanth model in a letter to Fricke:

"About 30 years ago, I was in the library of the Oceanographic Museum in Monaco, when a young man entered and asked to the present people if they believed he could sell to the museum a model of an articulated silver plated fish. He showed us the object, which reached (more or less) one meter (3.3 feet) in length. I recognized at once a very well represented coelacanth."

Unfortunately, the silver model was not purchased. Subsequent attempts by Hans Fricke and Jorg Keller in 1988 to locate additional silver coelacanth icons in Spanish churches, chapels, and local antique shops have so far proved fruitless. The two came across many silver fish containers and vessels for spices, tobacco and opium—all of which were still being manufactured locally, but no coelacanths. Fricke and Keller noted that these modern silver trade vessels were pale imitations compared to the skilled craftsmanship evident in the silver coelacanth which had been obtained by Steinert.

Amongst coelacanth enthusiasts, the search for more silver models continues. Only a century or two ago, examples of silver vessels were in great demand, especially by nobles of Spain, Germany, Sweden, and Holland, and it is possible that some silver collections may house a "strange" fish" carefully crafted in silver that is still awaiting discovery.

Madrid silver art expert, Professor Valdavinus, believes that at least some of the silver coelacanths came from colonial Mexico and later found their way to Spain. Spanish silversmith expert Rafael Munoa also proposed a New World origin of the silver models. Munoa also offered his opinion that in all probability the Steinert silver coelacanth was of American rather than Spanish origin (Plante, 1994).

Of course, this possibility raises an important new question: if these silver icons were handcrafted by artists in Mexico, or from early Spanish settlements along the Gulf Coast of Florida and Texas, what local fish of these areas were used as models? That is, were coelacanth-like fish used as models or were the silver vessels based on an American cousin of the coelacanth?

An Indian Coelacanth Population?

A 19th century Indian miniature originating from the Luchnow area may have been based upon a specimen captured in the early 18th century. The painting shows a Moslem saint, Hodja Hadir, standing on a fish with characteristic coelacanth features. The first dorsal fin is pushed somewhat forward compared to a real coelacanth, but this may simply be an artistic endeavor to make room for the saint on the back of the fish. If this miniature painting is a true likeness of a coelacanth then the question arises as to where the model for the fish depicted in the painting came from. Was a stray coelacanth caught off the Indian coast by fishermen and presented as a tribute? It is also possible that the painting is based on a specimen obtained by Indian seafarers. Since the 10th century, these seafarers ventured as far east as the islands of Indonesia and south and west to South Africa waters where they founded settlements in Mozambique. They undoubtedly visited the Comoros Islands, providing the situation in which a specimen, or at least a portrait of a coelacanth could have arrived in the court of the Moguls. That unfamiliar and spectacular animals were brought from islands to the Mogul nobles is supported by a Mogul-miniature of a dodo (circa 1625) now in the Hermitage (Brentjes, 1972). Of course, Indian fisherman also occasionally fished the long coastlines of the East Indies islands. Is it too far-fetched to suggest that the coelacanth which the Sulwesians called the King of Fish was presented to the kingly nobles of India? We think that it is at least a plausible scenario.

Coelacanth Cousins in the Gulf of Mexico?

The specter of an American coelacanth cousin was raised as early as 1953 in a *Science Newsletter* article based upon a fish scale sent to the U.S. National Museum in Washington. In 1949, ichthyologist Dr. Isaac Ginsberg received from a Tampa, Florida, woman an unusual fish scale she had purchased to fashion ornaments for the tourist trade. The scale was about 1 1/2 inches in width and "tarpon-like" but of an entirely different structure from that of the well known game fish. None of the fish specialists at the museum could identify the scale, so Ginsburg wrote to the woman asking for more scales and information but his letters were never answered. Ginsburg offered his comments about the unusual fish scale:

"This scale is like no other fish scale I have ever
seen. It is not the scale of any of the several hundred known fish species of the Gulf of Mexico, and
is apparently of primitive structure."

Ginsberg did not rule out the possibility that the controver-
sial fish scale was a coelacanth scale. Nevertheless—coelacanth
or not—the scales indicated the existence of some type of fish
"beneath American waters unseen and unknown to science."

In response to inquires by Gary Mangaicopra directed to the
Smithsonian Museum, Dr. Richard Vari (1988), Smithsonian's
Curator-in-charge of the Division of Fishes, established that
they did not have any records pertaining to that fish scale. He
further noted that Dr. Ginsberg was a member of the U.S. Fish
and Wildlife laboratory which was associated with the
Smithsonian—but not actually a staff member of the
Smithsonian.

Dr. George Zug (1988), Smithsonian's Curator of Division
of Amphibians and Reptiles added the following comments:

"Some of the older fish division staff do remember it and suggest that there is a possibility
that it may have been returned to the sender."

During the 1950's, naturalist, writer, and famed
cryptozoologist Wiley Ley (1959) summed up the circumstances
surrounding this controversy:

"Only one thing is known therefore, namely,
that a primitive fish, possibly a coelacanth, must
live off the Florida coast, most in the Gulf. One
day the story of *Latimeria* may have still another
chapter, this time one involving American waters."

Ley proved prophetic, for additional reports of unusual fish
and unusual fish scales were yet to come—all from the general
region of the Gulf of Mexico and Latin America.

Sterling Lanier, an American naturalist, book editor, and one-time science fiction author, told J. Richard Greenwell that he had seen coelacanth-like fish scales in 1973. Lanier was selling several of his brass figurines of fossil animals at an outdoor art show at a flea market along the Gulf Coast of Florida. Another artist was displaying a fish scale necklace that resembled those of the coelacanth. Lanier examined and sketched the scales (the original notes and sketches have since been lost) but did not buy the necklace. Lanier was told that the fish scales have been acquired from a shrimp boat "trash" pile (Greenwell, 1994).

New discoveries continued to spur interest in the existence of a Gulf of Mexico coelacanth population. In 1992, French naturalist Roland Heu came across a number of unidentified oval-shaped scales in a souvenir shop in Biloxi, Mississippi. According to Heu:

> "The part which would be visible in the living
> fish represents only the third of the length, so that
> the animal would be covered with a curiass of
> three supposed layers of very hard scales of a pe-
> culiar ossified structure."

French ichthyologist Professor Hureau of the Museum of National d'Historie Naturelle in Paris stated of the discovery that "should they come from Comoros, I would say that they are coelacanth scales." Other scientists, however, suggested that the unusual fish scales actually came from a giant Amazonian freshwater fish, the pirarucu (*Arapaima gigas*).

If all of these separate incidents are actually *Arapaima* scales, then the question naturally arises as to why they appeared in markets all along the Gulf Coast fish markets in Mississippi in 1992 and possibly Florida in 1949 and again circa 1973? Does this species also occur in the Gulf of Mexico, or are these scales simply harvested for export from the Amazonian region of Latin America?

Rumors of American Coelacanth Captures?

Prior to the mid-1950's, coelacanth pioneer J. L. B. Smith received a letter from a woman in Bermuda who was positive that a local fisherman had tried to sell her a coelacanth (Smith, 1956). Although Smith considered the possibility highly unlikely, he did note that the known ecological habitant requirements of the Comoros coelacanths are also found among the deeper reefs of the Caribbean and off the Bermuda coast. Of course, Smith was reluctant to speculate further regarding the possibility that the warm waters of the Gulf of Mexico might harbor hidden coelacanth populations.

In 1949 (or 1948?), a "four-legged" fish was caught off the Florida coast near Tallahassee, on January 18. This specimen was described as being some four feet in length with "fleshy fins" that somewhat resembled legs and were about three inches in length. The Florida Wildlife Association did not identify the specimen. The length of this catch is about right for a coelacanth but the "legs" are rather small. Could this fish have been an example of a misplaced lungfish? The South American lungfish, *Lepidosiren paradoxa*, grows to about 4 feet in length. It has thread-like pectoral fins, somewhat feather-like pelvic fins, and three pairs of holobranchs. Unfortunately, records of this specimen have long been lost and apparently no photographs were ever taken.

Cryptozoologist William Corliss (1955) spent some time in the southern coastal areas trying to find information about possible coelacanths in the Gulf of Mexico, only to come away disenchanted. In his words:

South American Lungfish

"... a year ago I spent 12 days at the Marine
Biology Laboratory of Hofstra University and St.
Ann's Bay, Jamaica. We talked a bit about crypto-
zoology, and no one mentioned coelacanths."

Speculative Conclusions

At the present time, the information available regarding "out
of place" coelacanths in the Indian, South Atlantic, North Atlantic,
Mediterranean Sea—and even the Gulf of Mexico—are based
wholly upon verbal reports, silver icons—and scattered fish
scales that may or may not have originated from coelacanths.

Fish scales from Florida and the Gulf of Mexico continue to
add dimensions to the mystery. The northern part of the Gulf
of Mexico has a sandy or muddy bottom, mostly shallow and
lacking the depths apparently preferred by coelacanths. On a
more positive note, the rocks and submarine grottos of ancient
coral reefs along the Florida keys and adjacent to a number of
Caribbean islands would provide good coelacanth habitat. The
Florida region may looked upon with suspicion as an area that
could in theory harbor an unknown species of crossopterygian.

While all of these conclusions are highly speculative, one
thing about coelacanths is certain. The circumstances attending the discovery of the first coelacanth population in 1938 and
the second 47 years later in an entirely different area of the
world, not less than 10,000 kilometers distant, suggests to us,
at least, that many surprises still await us regarding this remarkable living fossil.

Acknowledgements

The authors would like to give thanks to the following individuals whose generosity of data enable the
completion of this article: Dr. Hans Fricke, of the Max
Planck Institut fur Verhalten Physiologie, Seeweisen,

Germany; Dr. Burchard Brentjes (retired zoologist), Berlin, Germany; Dr. Raphael Plante, of Station Marine d'Endoume, Marseille, France; Dr. Roland Heu, of Paris, France; Dr. Maurice Steinert, of Bruxelles, Belique; C. Carpine, of Monaco, France; Dr. George R. Zug, Curator of Division of Amphibians and Reptiles, Smithsonian Institution, Washington, D.C.; Hjalmar Thesen, of Knysna, South Africa; Steven Kredel (German translator), of Greensburg, Pennsylvania. Coelacanth and lungfish sketches by Aleksander Lovcanski.

Selected Coelacanth Literature:

Alexandre, Michel. 1994. Le Coelacanthe. *Tele 7 Jours*, No. 1791, 14-30 Septembre : 28-129.

Anonymous, 1950. *Doubt*. No. 27 : 411.

Anthony, Jean. 1976. *Operation Coelacanthe*. Paris, Arthaud.

Bell, Shirley. 1967. Coelacanths A Century Ago? *Field and Tide*. 9: 8-11.

Brentjes, B. 1972. Eine Vor-Entdeckung des Quasten-flossers in Indien? *Naturw Rdsch* 25: 312-313.

Browne, Malcolm W. 1988. Are Scientist a Threat to Rare 'Fossil Fish'? *New York Times*, March 2, 1988.

Bruton, Michael. 1985. The Silver Coelacanth. *Ichthos*, No. 8 (March): 16.

Bruton, Michael. 1989. Is There a Madagascar Coelacanth? *Ichthos*, 23: 25.

Bruton, Michael. 1989. Does the Coelacanth Occur in the Eastern Cape? *The Naturalist* 33: 5-13.

Bruton, M, M. N., A.J.P.Cabral, and H. W. Fricke. 1992. First Capture of a Coelacanth *Latimeria chalumnae* (Pices, Latimeriidae) off Mozambicque. *South African Journal of Science* 88: 225-227.

DeSylva, Donald P. 1966. Mystery of the Silver Coelacanth. *Sea Frontiers* 12: 172-175.

Erdmann, M. V., R. L. Caldwell, and M.K. Moosa. 1998. Indo-nesian "King of the Sea" Discovered. *Nature* 395: 335.

Forey, P. 1989. Le coelacanthe. *La recherché* 215: 1319-1326.

Fricke, Hans. 1989 Auf Spurensuche Quastie im Baskeland? *Tauchen* 10: 64-67.

Fricke, H. W., and K. Hissman. 1990. Natural habitat of the Coelacanth. *Nature* 346: 323-324.

Fricke, H. W., and K. Hissman. 1994. Home range and migrations of the living coelacanth *Latimeria chalumnae. Mar. Biol.* 120: 171-180,

Fricke, H. W. And K. Hissman. 1992. Locomotion, fin coordination and body of the living coelacanth *Latimeria chalumnae. Env. Biology Fish* 34: 329-356.

Fricke, H. W., and R. Plante. 2001. Silver coelacanths from Spain are not proofs of a Pre-scientific discovery. *Environmental Biology of Fishes* 61: 461-463.

Greenwell, J. Richard. 1994. Prehistoric Fishing. *BBC Wildlife* 14: 33.

Heemstra, P. C., A.L. J. Freeman, H.Y. Wong, D. A. Hensley, and H.D. Rabesandratana. 1996. First authentic capture of the coelacanth, *Latimeria chalumnae* (Pices: Latimeriidae) off Madagascar. *South African Journal of Science* 92: 150-151.

Holder, M.T., M. V. Erdmann, T. P. Wilcox, R. L. Caldwell, and D. M. Hillis. 1999. Two living species of coelacanths. *Proceedings of the National Academy of Science* 96: 12616-12620.

Kredel, Steven. 1994. Personal communication to Gary S. Mangiacopra.

Mouton, Patrick. 1990. A La Decouverte des Derniers Coelacanthes. *Le Monde de la Mer* 49: 32-39.

Nelson, Joseph S. 1976. *Fishes Of The World.* John Wiley and Sons : N.Y., London, Sydney, and Toronto.

Plante, R. 1997. Un fossile vivant en sursis? *Oceanorama* 27: 16-18.

Schlieven, U., H. Fricke, M Scharti, J.T. Epplen, and S. Paabo. 1993. Which Home for Coelacanth. *Nature* 363: 405.

Smith, J.B.L.1953. A Coelacanth in Kenya Waters? *The South African Angler* 8: 19.

Thomson, Keith S. 1991. *Living Fossil—The Story of The Coelacanth.* W.W. Norton and Company: New York, London.

A Mysterious Kentucky Water Cryptid

Barton M. Nunnelly

On a sunny morning back in the late 1980s, two teenage boys approached the muddy backwaters of the swollen Ohio River in Henderson County, KY. As they walked down Klondike Road, they loaded the small .22 caliber pistol they had brought along with the intentions of getting a little target practice in by shooting turtles off logs. They reached the point where the mighty Ohio flowed over the road and stopped, searching anxiously for possible targets—these were plentiful. Andy A. went first. His friend, Mike B., had persuaded his father to drive them to the isolated spot in the Geneva River bottoms, but his father had chosen to remain behind in the car to read the daily newspaper. He wasn't far away and could hear the shots ring out clearly as the two boys took their turns. After shooting for a few minutes, the boys had paused when Andy saw something 'bob up' from the water just a few yards from where the two stood. It was a green colored, semi-circular 'hump,' like an auto tire, covered in moss with dark green circles about the size of baseballs spaced about a foot apart. He called his friend's attention to it and they both were amazed when another hump surfaced close to the first one. Then another, and another. All they could figure was that someone had tied a string of tires together and they had now come loose in the flood. All together, five 'humps' surfaced and they were getting a feeling that something wasn't quite right about the thing. All the 'humps' were segmented and looked a great deal like the skin of a lizard. Then, at the front of the 'humps,' a head arose from the water and the youths froze in their tracks, mouths dropping agape in wonder and shock. They could clearly see its elongated snout and dark black eyes. It slowly swung its head in their direction and, upon seeing them, almost as if surprised by their presence, the creature surged forward and sank very quickly, disappearing once again beneath the swift, muddy water.

I interviewed one of the witnesses, Andy A., in 1999 and found his story to be all too credible. He claimed the creature was over 30-foot long, as big around as a tire, with no scales. He also described a darker line which ran length-wise down the middle of its side, resembling the coloration which can be found on gamefish such as bluegill and bass. When it fled, he claimed, it did not dive head first into the water, but 'sank down' extremely fast. It was gone in a couple of seconds and swam by undulating its body from side to side like a huge snake. I was able to produce a sketch, under his directions, of this curious cryptid. During a follow-up interview in 2005, Andy gave me the exact same details as he did in the '99 interview. He was unable to add anything to the account or take anything away from it, which speaks much about the validity of his claims. The other witness, Mike B., just as he has done from the beginning, still refuses to talk about the event.

Such a claim, that animals of an unknown and unclassified nature are to be found in the Ohio River, is not altogether unique or extraordinary. Many people have reported much the same down through the years. I myself am not unfamiliar with such animals, having seen on a couple of occasions, and in the company of witnesses, a similar creature in the Ohio River some twenty miles away in Stanley County. Although viewed from a greater distance, the animal was similar to the Geneva cryptid in that it appeared serpentine except for the duck-like bill, or beak, on the front of its head. Distance from the sighted animal was roughly 50 yards and its appearance was that of a younger and smaller animal. The exposed portions were comparable in size to a grown man's forearm jutting from the water's surface. Many factors can sometimes combine to complicate sightings of this nature, most common being distance and the uneven quality of the river's surface. Some would suggest that it was merely a large duck, but my wife and I watched this creature for almost thirty minutes and it did not behave in an explainable manner. It would surface, then look around several times and sink back beneath the waves, only to come up again twenty yards away. It repeated this routine until it finally failed to resurface completely. At no time was a body visible above the water. Only the neck and head.

At a summer camp out in 2001, James K. reportedly saw an even smaller version of the exact same thing as it swam close to shore for nearly two hours. It was also described as being snake-like, with a beak which resembled a duck's. It was only about three feet long. This took place within two miles of the Geneva sighting. Also from this general area I am aware of another sighting of a monstrous aquatic unknown. It happened in the late 1970s beneath the twin bridges which span the Ohio and connect Henderson, KY, to Evansville, IN. Back then there was a floating gas station anchored in the area which provided fuel for the numerous boaters who plied the river both for business and pleasure. Randy O., an attendant, was working one night when, he claimed, a huge, terrifying creature rose up from the river just a few yards away. It scared him so badly that he immediately quit his job and never returned. He is now a Baptist minister and is unwilling to give further details about the physical description of the creature.

The location of the Geneva sighting is included in The Sloughs Wildlife Management Area, a ten-thousand-acre strip of wetlands, forests, and well cultivated fields which is bordered on the North by the Ohio River. All along the Kentucky side of the river the landscape remains unchanged for hundreds, if not thousands, of square miles. The area is very sparsely populated, with long, lonely stretches of bottom land between one dwelling and the next. As a consequence, the odds of seeing the emergence

Author's Reconstruction

of any aquatic animal other than turtles and the occasional fish are very remote. Nevertheless, sightings of these animals continue to happen again and again.

Interestingly, only a few miles from where most of these water cryptid sightings took place, the Green River empties into the Ohio and could be the source from which many of these unknown aquatic animals originate. The Green River is one of the deepest rivers in the world, second only to South America's Amazon River. Long stretches of this waterway are considered bottomless and it is the only north-flowing river in the entire United States. Many strange things have been seen in this river as well. Once, in 1998, as I was traveling east over the Spottsville bridge, which spans the Green river, I happened to notice a disturbance in the water below. It was just before dawn and the light was not conducive in revealing much detail, but I could see two slender objects, each about 15- to 20-foot long, apparently engaged in the act of fighting each other. They were creating large waves as they splashed about and I merely assumed them to be more examples of the giant catfish which are well known locally to exist there. It has since occurred to me that a twenty foot long example of such a specimen most probably would not be slender in appearance, so the identity of the animals involved must remain a mystery. However, a cryptid sighting occurred in this same area in the early 1980s and involved a fisherman who was standing just above the lochs when a creature emerged from the deep channel just offshore. It was described as being long and slender, of immense proportions and had two eye stalks growing from the top of its head much like a snails. The fisherman fled, with all due haste, as the thing began to slowly approach.

The actual identity of these aquatic anomalies may never be discovered, but one thing is sure. Of the nearly two-hundred varieties of fish which are known to inhabit Kentucky's rivers and streams... these creatures are not among them, but something else entirely.

Didi of the Imagination

Chad Arment

Investigators who search through old newspaper archives (whether microfilm or digital) looking for cryptozoological accounts will sometimes come across that rare but scintillating story of a vivid encounter with an undescribed beast. And story it is—not just dry factual accounts of a shadowy brief encounter, but a narration that excites the blood or even tugs at the heartstrings. A story like this:

Spirit of the Woods.

It Is Said to Haunt the Dense Forests of Guiana.
The Indians Are as Afraid of Didi as of Death—An
American's Adventure in a Tropical Wilderness—
Escape of an Ugly Monster.

[Special Panama (C. A.) Letter.]
There is one experience to which, on account of the terrible significance that tradition attaches to it, the Indian of the Guianese forests has never become reconciled. To all else his stolidity, that equanamity of unconscious fatalism which is his most

59

distinguishing characteristic, is invulnerable; but this one exception that "proves the rule" has the power to set his iron nerves a-trembling and makes a driveling poltroon of the boldest. It is the voice of the Didi, the Evil Spirit of the Woods, and is the signal that he is visibly abroad seeking whom he may destroy, for, like the rattlesnake, the demon may not approach his human victim without giving this timely warning. Once heard there is no mistaking the sound of the Didi's voice, since it is totally unlike anything else heard in the forest. It is a prolonged melancholy whistle, beginning abruptly as a locomotive's toot in a high key and dwindling down to the merest thread of sound. The sight of this being is supposed to be instant death, for from his eyes shoot forth flames that blast and reduce to cinders the luckless mortal on whose vision this hellish apparition looms through the darkness of the night. Hence, when that piercing cry goes echoing through the forest, every Indian hurriedly wraps his blanket about his head and remains thus muffled, preferring the risk of asphyxia to exposing his eyes to the horrific presence, until the light of dawn drives the Didi back to the nether world.

Although no one has ever seen and lived to describe the monster, tradition gives it the form of a gigantic ape, larger than a man and covered with a matted mass of fiery red hair. Of course the superstition is absurd beyond consideration, but apart from the elements of the supernatural it is thought by many that it has some foundation in fact; that notwithstanding the pronounced skepticism of men of science some still unknown species of simia exists in those forests which avoids the vicinity of man and roams abroad seeking its prey only by night and in the farthest and darkest recesses of the woods. And in the light of a most thrilling adventure that once befell me I am constrained to share that belief. During frequent expeditions into the interior of the Guianas I had grown familiar with the voice of the Didi and had a contract with myself to trace it to its source, but the opportunity was never favorable. Time and again had I been roused from my sleep at the dead hour of midnight during a howling storm to renew that contract and then quench my irritation in a burst of genuine merriment at the grotesque figures cut by the Indians, as in feverish haste they tumbled over each other out of their hammocks and wound fold on fold of suffocating blanket about their heads, grunting in dolorous concert the while there sounded loud and shrill above the howl of wind and lash of rain the truely demoniac whistle of the Didi. However, all things come to those who wait, and my opportunity came at length with an experience the like of which I should not care to undergo again.

One night, when encamped on the banks of the Caroni river, I awoke with a start to find my Indians bunched in a tangle of legs and arms, convulsively struggling amid fluttering folds of blanket and tumultuous waves of hammock, yelling: "Didi! Didi da, come!" Their terror was unusually acute, and no wonder, for the air was full of the voice of the Didi. It came, not from the far recesses of the woods, but apparently from somewhere in the immediate vicinity of the camp, and the demon might reveal himself at any moment. There was no storm on tonight, and the full moon rode high in a cloudless sky, flooding river and forest with here tropical radiance. Now or never should I identify the Didi, and with contemptuous disregard of the Indians' warnings I was soon on his trail, following the whistle through a darkling glade of the forest down which it receded when I appeared.

Now and then I thought a darker darkness damasked itself on the shadows that filled the spaces beneath the heavy canopied trees, but nothing could be distinctly seen there. The whistle, however, never ceased for more than a few seconds, and I was sure that the chase was not increasing the distance between us. The trail led away into the heart of the forest, but suddenly turned abruptly back toward the river until I could catch the gleam of the water in radiant patches between the foliage. This continued for some time longer, and many miles must have slipped under my eager feet, when a flood of light broke ahead and in another moment I found myself on the edge of an extensive opening on the river shore. Across this, and in full view under the moonlight, raced a gigantic, a monstrous creature like a gorilla, which I judged to be at least eight feet in height. He ran, or rather lumbered along, upright on his hind limbs, swinging in his left hand a formidable looking club. The body was covered with hair, and the face was the most unutterably hideous I have ever beheld. I had an excellent view of him as he stood for a moment and peered apprehensively over his shoulder at me.

Sighting to hit him in the spine, I fired. The incessant whistle, which he had just begun, instead of dwindling reversed the process, as it were, and waxed higher and higher until it attained a frightful compass and volume! And then—. But where was the Didi? Amazed, scarcely crediting the evidence of my senses, I ran forward to within a few feet of the spot where he had just stood, and halted literally paralyzed by a superstitious dread that took possession of me and overwhelmed calm reason. The monstrous beast, or rather thing, had *disappeared*—mysteriously vanished, that is, for under the circumstances it was not possible that he could have regained the cover of the forest! Was it all a dream? Was I the sport of a nightmare? No, for there sailed the moon, sloping to the west above the forest across the river, whilst about me murmured the restless symphony of the forest, and the firm earth answered: "Yea, verily," to the query of my stamping feet. Then my knees smote together, and the parched tongue clove to its burning roof. The very loneliness of the wilderness in which I so delighted, and that contributed a perennial charm to the life that I led face to face with nature and, as I was wont to hope, with nature's God, now drove me to the verge of frantic madness. For, lost to the world and buried amidst that desolation, I had become the sport of malicious evil spirits!

Pray? I was too distraught to even think of that. Indeed, after that fierce bout of stamping—stamping on the earth as if to question her through my sense of feeling as to the reality of all things—all remains a blank until I found myself back at the camp where the pale flush of dawn stealing through the trees tinted the cold ashes of the fire, and roused the Indians who were cautiously unwinding their head wrappings. Then only did reason fully assert herself. But with daylight and companionship courage returned, and feeling heartily ashamed of my scare I determined to thoroughly investigate the mystery: moreover, I had dropped my Winchester which must be recovered.

After a bath and substantial breakfast I felt ready to face a legion of airy demons, and accompanied by two Indians, who readily detected my trail, I returned to the scene of that terrible adventure.

There could be no mistake, for there lay the rifle glistening amid the short grass. And a little farther on lay the solution of the mystery. How absurdly simple it all seemed in the broad light of day! A ragged hole in the ground, about which lay a scattered debris of wicker work, tufts of grass and loose mold, revealed the secret of the disappearing Didi. In my excitement I must have aimed wide and but slightly wounded the gorilla—or whatever else the creature happened to be—and he had fallen into one of those automatic tiger pits that the Indians plant in open spaces near the rivers, where those beasts love to gambol on clear nights after having secured their prey. Thus had the chase escaped me; and after I left, taking advantage of his great height and immense strength, he had reached up and torn away the trap roof, easily freeing himself.

It was truly humiliating to have thus lost an opportunity of capturing a Did, but at the moment of solving the mystery I was so full of thankfulness at having, as if by a miracle, escaped following him into the pit, that I gave but scant heed to the loss. But I have never ceased to regret the ridiculous panic that snatched from me to reserve for another the actual discover and naming of the South American chimpanzee, of the existence of which I am, of course, convinced.

T. P. Porter

—Davenport (Iowa) *Weekly Leader*, Sept. 13, 1893.

Of course, this is fiction. Specifically, it is one of the earliest forms of science fiction, the newspaper tale that often used mechanical wonders or natural mysteries to spin a story for entertainment under the guise, "There are more things in heaven and earth, Horatio..." Newspapers had to fill space in the late 1800s and early 1900s, and editors at some newspapers were more than willing to include fabricated tales (which were then reprinted by smaller newspapers). In this case, T. P. Porter wrote several stories of Central and South American adventure published in the late 1890s: surviving a tropical earthquake, fending off bandits, fighting Cuban insurgents, etc.

I previously noted (in *Cryptozoology: Science and Speculation*, 2004), that the supposed story of a West Virginia roc ("A Modern Roc," St. Louis (MO) *Daily Globe-Democrat*, Feb. 24, 1895) is a newspaper hoax. But how do we distinguish between false stories and legitimate sighting reports? For the West Virginia story, I was able to visit an extensive archive of newspapers in that state and show that no such story originated there—it was fabricated by the St. Louis newspaper staff. That newspaper was well-known for making up stories of scientific marvels, which is a good reason to be wary of any story coming from its pages.

Many of these stories are, in fact, just that. They are extended narrations with at least a simplistic plot. They do not merely stick to the "facts." Many early short story writers, like Jack London and Samuel Clemens, honed their craft working at newspapers. They included quite detailed scenarios with names and places, much of which could never be corroborated. Details do not always betray a fictional foundation, but caution is warranted when it looks like the writer has tried to prop up the story with too much unconfirmable data.

Here is another example of newspaper cryptofiction, dug up by Gary Mangiacopra:

Found a Real Merman

Pacific Porpoise Hunters Kill a Queer Marine Monster
Scarred In Many Battles
Almost Human in Appearance from the Waist Up,
But Has the Tail of a Porpoise —

Tacoma, Wash., Oct. 31, 1896

What appears to be a genuine merman was brought into this port last week by a party of Englishmen. They had been porpoise fishing in the Pacific and were more than confounded at the extraordinary creature they captured. They came in with their prize fully convinced that the old stories about merman and mermaids were all true in spite of the scoffers. The man who deserves the credit of this wonderful discovery is Maj. W. E. Thorncliff, of the English army. The major was at first rather averse to giving the details of his novel adventure, fearing that he would be classed with the spinners of ordinary fish yams of Puget Sound, but knowing that his social and official position put his word above question, he finally consented to relate his unique experience, and to exhibit his interesting captive-only stipulating that I should repeat the facts exactly as he stated them, and describe the sea monster precisely as it really is.

This is the story of the Major's adventure, in his own words:

"Our party, which consisted of several English noblemen, a French statesman and a Russian prince, left Hokondach, Japan, on a fishing and hunting expedition to this coast, on board of Prince Gerenoff's steam yacht Anedamoff, on June 20, and we cruised along the shores of the Aleutian Peninsula, calling into many very fine bays and harbors along the coast.

"We shot on shore and fished in the waters of both Behring Sea and the Pacific Ocean, and have, as trophies of our skill, a fine collection of pelts, as well as skeletons of many rare creatures.

"But the climax of all came on the morning of July 26, when we were off the Island of Watmoff.

Our men sighted a school of porpoises, among which could be seen several white ones.

How It Was Caught

"Our hunting boat was lowered, and Lord Devonshire, the Earl and I, with the boat's crew, put off from the yacht, determined to capture some of the rare sea pigs. After pulling about four miles we found ourselves in the center of the school, and Lord Devonshire got a shot at one of the white boys with a large express rifle, which quickly ended its career.

"Just as we were putting our guns away the Earl called out, 'See that!' pointing to a most startling looking beast, not more than a cable's length away. Picking up his express he fired point blank at it, striking the creature between the eyes. The shot, though it did not kill it, so stunned it that it lay perfectly still on the surface of the sea.

"As our boat hauled alongside we saw the most hideous and uncanny looking monster probably that human eyes ever looked upon. Although at a distance it might perhaps be mistaken for a porpoise, as we came near we saw that it could truly be described by no other name than that of 'merman.'

"As we reached over the side of the boat to haul the creature in it regained some of its vitality. It caught the boat by the gunwale amidships and had it not been for the fact that when the arms came up out of the sea we naturally shrank to the other side of the boat it would, without doubt, have capsized us. One of the men picked up an ax and quickly dispatched the monster.

"The better way now would be for you to come with me and I will show you the strange creature which I am now taking to England to present to the British Museum. After seeing it you will, I am sure, be inclined to the opinion that once it is placed there it will easily outrank all of the many strange things to be found in that great repository of the world's rarities."

Then the major led the way to a store-room on Pacific avenue, where, in the middle of the floor, was a large, coffin-shaped box. It was ten feet long, three feet wide, and three feet deep. Taking a screwdriver, the Major unfastened the top. All that could be seen was some ice, covered with a white woolen blanket. Taking the blanket by the end he quickly removed it, and as he did so the sight of the contents of the box almost froze my blood, for right before my eyes was apparently the naked body of a large man.

Hero of Many Fights

The Major then removed the cloth which covered the lower part of the body. This is exactly the same as that of an ordinary porpoise. The merman is one of the most remarkable freaks nature ever put together. The strange monstrosity measures ten feet from its nose to the end of its fluke-shaped tail, and the girth of its human-shaped body was just six feet.

It would weigh, it is estimated, close to 500 pounds. From about the breast bone to a point about where the base of the stomach would be, were it human, it looked exactly like a man. Its arms, quite human in shape and form, are very long and covered completely with long, coarse, dark reddish hair, as is the whole body.

It had, or did have at one time, four fingers and a thumb on each hand, almost human in shape, except that in place of finger nails there were long, slender claws. But in days probably long gone by it had evidently fought some monster that had got the best of it, for the forefinger of the right hand, the little finger of the left and the left thumb are missing entirely. Immediately under the right breast is a broad, ugly looking scar, which looked as if some time in the past it had been inflicted by a swordfish. On the sides and body of the monster are numerous other evidences that its life in the ocean had been far from a placid one. There is hardly a space the size of one's hand that does not show evidence of having at some time or other received wounds.

In Britain's Museum

When the hideous body reaches England that country can safely say that it possesses the strangest freak the mysterious waters of the Pacific ever gave up.

"Now mind," was Maj. Thorncliff's parting salutation, "don't in any way try to embellish what you have seen and heard, but just tell the plain facts, and though this coast may be renowned for strange and weird stories, this story of the merman, simply and truthfully told, will, I am confident, prove the adage, 'Truth is stranger than fiction.'"

Now it only remains for some man as responsible and well known as Maj. Thorncliff to discover the mate of this merman, and we will be convinced that the old mariners had not, after all, the wonderful powers of imagination and romance which we have so long ascribed to them.

—Washington (D. C.) *Post*, Nov. 1, 1896.

So, when perusing old newspapers, take nothing for granted. The mystery behind zoological discovery is strong enough for fiction writers to incorporate cryptozoology into their writings. And, that in itself provides us opportunity to see how cryptozoology influences culture, and vice versa.

I'll continue looking for more examples of these stories. They may not be as polished as traditionally published cryptofiction, but they enjoy a certain enthusiasm for nature's spectacles that tweaks the dogmatism rampant in modern science.

Freshwater Cephalopods

Gary A. Mangiacopra
Dwight G .Smith

"Mr. Bond, they have a saying in Chicago: 'Once is happenstance, twice is coincidence, the third time it's enemy action.'"

—Arch-villain Auric Goldfinger to secret agent James Bond in Ian Fleming's classic spy novel, *Goldfinger*.

"Today's fishing report from Lake Conway: Bass fair, crappie very good, bream good, catfish fairly good, octopus excellent"

—Arkansas News Bureau, 2003

To those familiar with the fledging and slowly growing field of cryptozoology—the study of hidden animals—one invariably thinks of investigators seeking large and exotic animals in the four corners of the globe: Nessie of Scotland, Bigfoot in North America, Mokele-Mbembe of the African Congo, to name just a

71

few that have graced headlines of newspapers and magazines of many nations and of which the general public is aware.

There is, however, another aspect of cryptozoology that is greatly overlooked, and that is the growing problem of what we are terming out-of-place-animals. The present article is concerned with the ever increasing number of animals (and plants also) that have been reported and in some cases captured far beyond their normal range, in non-native regions where they should never have been in the first place!

Out-of-place-animals are frequent headliners. In the United States alone, non-native animals such as Nile monitor lizards have been captured in Florida (1,2,3) along with walking catfish, giant South American toads, (4, 5) and a growing assortment of exotic parrots, parakeets, and other pet trade birds.

The recurrent and popular urban legend that New York City sewers are infested with alligators is in reality not quite an urban legend as alligators have been caught in this city, along with 50-pound giant snapping turtles. (6,7, 8, 9) Even the Nutmeg State of Connecticut has recorded captures of alligators over the past decades. (10, 11, 12, 13, 14, 15, 16) In the year 2000, a Kennebunkport, Maine resident shot a 2 1/2 foot alligator that nipped at his pants! (17)

There have even been reports of wild Australian kangaroos in the United States and even as far north as Canada. For example, residents around Lochabere Lake, Nova Scotia, repeatedly reported seeing a kangaroo-like animal that left footprints for several years during the mid-1980s. (18)

These incidents are just a random sampling of out-of-place animals being seen and even photographed on video tape and leaving tangible tracks that conclusively documents that they were there, where they should not be. The usual explanation for most observations of these out-of-place-animals is that they were once exotic pets that had been obtained legally and once the reality of the expense, size, and personal danger of such pets become too much to maintain, they lose their cuteness to their owners. They were simply let loose by the owners to fend for themselves— or perhaps they escaped and took up residence in the wild.

An equally serious problem are pets that were acquired illegally by their owners which accidentally escaped from their confines and subsequently adapted to the surrounding local wildlife areas, even to the point of breeding in the wild, producing a non-native population.

But how does one explain away reports and even of captures of wildlife in regions very far from their native specialized habitat which one would normally assume they could not have survived in at all? Such is a zoological case dealing with—of all species — cephalopods, species of squids and octopuses which have been observed and/or captured alive in a number of freshwater river systems of the United States! A zoological impossibility—but such accounts have been published in local newspapers and scientific journals. (19)

The 650 species of living cephalopods include two great groups, the Nautiloidea, which includes the world-famous *Nautilus* and its extinct relatives, and the Coleoidea which includes the cuttlefish, squid, and octopuses. All are exclusively marine. There is not one that is known to have adapted to freshwater environments. Of course, many occur in brackish tidal pools and some ascend brackish rivers, but never for any great distance. Generally the extent of the estuarine portion of a river is sufficient for them. So far as we are able to determine, the lowest salt content range tolerance which cephalopods survive ranges between 16-25 parts per thousand of salt by weight, about equal to brackish waters found near the mouth of rivers emptying into salt waters such as oceans, bays, or sounds.

The two cephalopod members of immediate interest to us— the squid and octopus—are predatory mollusks that in some cases can reach great size. Together they are considered to be the most advanced and possibly the most intelligent of the invertebrates.

The squid is the more spectacular and wide-ranging of the two, occurring from cold polar waters to the tropics. While most are pelagic or live near shore, the giant squid and certain of the octopuses are known to inhabit great depths. Of the two, the squid are long-bodied and adapted for relatively rapid movement as they hunt prey. Octopods are much more common in mid- to shallow water habitats and are a relatively common reef predator. However, a number of deep-water octopuses and squids exist, illustrating the fact that this group is both diverse and adaptable in habitat exploitation. (20, 21)

So the question becomes, why for the past century have we had a good number of documented reports of cephalopods—particularly octopuses, in freshwater habitats of the world? And why have so many of these published accounts been neglected and forgotten almost immediately after they appear in the news media? Like alligators in the sewers of New York or parakeets in a Boston backyard, most such accounts are largely forgotten, although we must confess that city sewer alligators continues to grip the media imagination, especially of Hollywood from time to time.

The following accounts of out-of-place cephalopods can be examined with a skeptical mind, but we ask that the reader also keep an open mind. Perhaps some of these accounts are nothing more than hoaxes. More likely, many of them represent discarded fish bait. Or could they be another phenomena? That is, could octopuses and squid survive in freshwater lakes and ponds for at least a while, again, in a manner reminiscent of city sewer alligators?

A New York Lake Squid

The earliest published account of an out-of-place cephalopod is from Onondaga Lake, New York, which lies near the center of the state, within the Onondaga Indian Reservation. Notice of the capture of a squid in the lake was published in Volume 16 of *Science* in the year 1902. The letter, published in the December issue of that scientific journal by Dr. John M. Clarke, related that a few days prior, the newspapers told the story of a Syracuse citizen, Mr. Terry, who while fishing for minnows in the shallow waters of the lake caught a strange-looking fish in his net. The specimen was brought to Syracuse University professor

John D. Wilson who identified the fish as a squid. Soon after-
wards, a second living specimen was caught in the same locale
by a local restaurant owner, Mr. Lang, whose place of business
was on the southeast corner of Lake Onondaga. Both squids
were alive when captured. The first specimen was boiled, then
preserved in alcohol. The second squid was retained in Mr.
Clarke's possession. Clarke added further comments regarding
his discovery of the squid. As to this being a "fake" done by the
work of a local hoaxer, Clarke questioned as to why choose this
lake and not some better known New York state lake. Professor
Wilson contributed the information that "there were several
hotels about the edge of the lake from which oyster and clam
shells are thrown into lake waters." On this, Clarke commented,
"...it hardly seems that this fact opens a possibility for the introduction by this means of the eggs of one of our Atlantic squids
into conditions that would permit their hatching." In other
words, Clarke thought it highly unlikely that the adult squid
captured represented squid that had hatched and then survived
to become adults in the waters of the lake.

Clarke ended his letter with this cautious statement, "Perhaps all will depend on the determination of the zoologist, the
specimen in my hands will be turned over to an examination by
an expert." (22)

The expert to whom Clarke forwarded the specimen was New
Jersey resident, Dr. A. E. Ortmann. Writing on the 12th of December from Princeton University, Dr. Ortmann gave his critical
analysis of the second squid specimen which ultimately appeared
in Volume 17 of *Science*. (23)

Dr. Ortmann used the descriptions of North Atlantic cephalopods given by Professor Addison Emery Verrill, and impor-
tantly, comparison with two well-preserved male and female
squid from Provincetown, Massachusetts, that were preserved
in the Princeton University animal collections, to verify the
identify of the species as a squid.

According to Dr. Ortmann, the 13-inch female specimen apparently belonged to the well-known species of short-finned
squid that was relatively common along the northern Atlantic

coast. The scientific name of this squid is *Illex illecebrosus* (Lesueur). Dr. Ortmann was also able to conclude that the female squid "was in no wise different from *Illex llecebriosus* of our northeastern coasts."

In the last paragraph of his letter to *Science*, Dr. Ortmann took a strong position towards the comment that former writer Clarke had noted to the effect that, "if it is a fact that this species lives in this lake, the only explanation is, as suggested, by a former, post-glacial connection of this lake with the Saint Lawrence Gulf."

Dr. Ortmann further appended his statements with, "but I am loathe to believe that this species lives in Onondaga Lake." He suggested an alternative and logical explanation that "...this squid is largely used for bait and the capture of squid forms a regular trade on our northeastern coasts. Could it not be possible that somebody has secured by purchase a barrel of squids, to be used as bait at the locality where our specimen was found?" (23)

These same statements were echoed over six decades later by an unnamed editor of *Sea Secrets* magazine (24) who related that dead squids had probably been bought by a bait dealer on the shore of Onondaga Lake, to be used by fisherman as bait. The bait dealer, or the fisherman, had thrown the excess bait overboard at the conclusion of the fishing venture, and they drifted near shore to be found and thereby make the pages of *Science*. The only thing wrong with this plausible scenario is, of course, that at least one of the squid was alive when first captured.

Ironically, a more thorough—and forgotten—explanation was given only a year later in 1903 by the United States Fish Commission, which by then had become involved in the great squid controversy. In a letter to the Washington *Post*, the chief of this division explained, "Few realize how many scientific hoaxes are run down by the United States Fish Commission." (25) and furthermore, "the recent findings of squid in Lake Onondaga may be the result of one of these practical jokes."

Going into further detail, the Fish Commission Chief gave additional examples of claims of fisherman capturing a variety of species of saltwater fish in the Great Lakes region, most of which were later determined to be simply fraud on the part of the tellers. And as for the Lake Onondaga squids:

"As I have already stated the overweening desire and inclination on the part of the ill-informed to hoodwink the government scientists, and those also of the universities, is so widespread and general that most people are astonished when they learn of its extent and frequency. There appears to be a great deal of this nonsense in western New York. Several years ago Prof. Hargitt of Syracuse obtained from persons who asserted that they had taken in Onondaga Lake a sargassum fish (*Pteraphyrne histrio*), and now the Rochester savants are stirred over the finding of a squid in the same body of water. The squid were sent to me for examination, and there is no doubt that they are the regular North Atlantic cuttlefish. Unless the whole thing is a hoax, I can readily imagine how they reached Onondaga Lake. The squid is an excellent bait fish. In Boston and New York they are sold to the masters of cod fishing vessels, who take along several thousand frozen in ice. I conjecture, therefore, that some fisherman purchased several blocks of squid in the New York market for use in fishing in Onondaga Lake, and having more than they required, left the remainder lying on the strand of land were the specimens are said to have been found."

Regretfully, after considering all of the various actual and possible scenarios and explanations, regarding the circumstances of the Onondaga Lake squids we must conclude that they almost certainly represent bait fish. And these out-of-place cephalopods are really examples of out-of-place bait fish. However, amongst all of the excitement of charge and countercharge,

Octopus

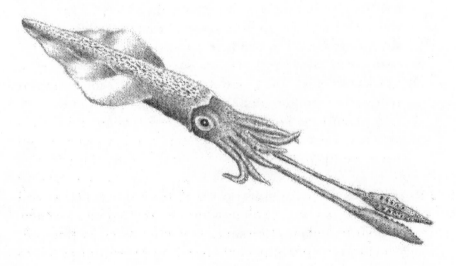

Squid

of all the concern about the possibility that they were planted as a hoax, no one seems to have taken noticed that at least one of the cephalopods was actually captured alive. This provides impetus to the possibility that eventually a few "fish bait discards"—if that is what they were—may someday survive under the right circumstances.

Freshwater Cephalopods of the Depression Years

The next round of reports of out-of-place cephalopods reported to have turned up in various freshwater habitats made news fillers during the rock-bottom years of the Great Depression.

As the economic depression hits its lowest point in 1933, a diverting Christmas Day release by the Associated Press from Charleston, West Virginia, added a bit of Christmas cheer and excitement. (26) On the day prior, the 24th of December, two Kanawha River fisherman found an unexpected intruder—a three foot long octopus—which flopped over the side of their row boat to sprawl untidely in the bottom. Robert Trice, who carried a crutch due to a recent automobile accident, impaled the creature while Ross Saunders plunged a knife into its gullet. Its response was to omit an inky spray. (27)

The freshwater octopus quickly gained local notoriety and, two days later, it was being considered as a possible exhibit at the state department of archives and history.

Four days later, however, the excitement ended and the offer of a state hall of fame collapsed when the local Charleston *Gazette* exposed this sham as nothing more than a Christmas day hoax. The local police sought Ross Saunders and another man for investigation in connection with the theft of a barrel of fish of which, it seems, also contained the octopus. It seems that the great fish barrel thief had presented Trice with the octopus, who in turn thought it would be a "swell Christmas joke" to say that he had killed it in the river. The larger question as to how this octopus got to West Virginia in the first place was answered

by city detectives John Wooster and E.N. Shuck, who found that
the barrel had been shipped from Boston to Mrs. Chris Racona,
owner of Slack Street Grocery Store. (27, 28,29) This, unfortunately, completely closed the case on this out-of-place cephalopod.

A Middle Oregonian Octopus

By mid-depression two more out-of-place reports of cephalopods were recorded. The first concerned a West Coast cephalopod said to have been captured in central Oregon. This capture intrigued several state biologists sufficiently to travel and
investigate the remarkable discovery. On the 15th of March, reports came from Redmond City (not far from the geographic
center of the state) that Jack McDaniels had caught an octopus
28 inches long in Metolius River. The catch was placed on display in the town's hardware store. This discovery prompted
Stanley G. Jewett, head of the United States Bureau of Biological Survey in Oregon to check out its authenticity and, if it
proved to be authentic, to preserve for posterity this unique
specimen. (30-34)

The question of its validity was challenged by William L.
Finley, one of Oregon's best known naturalists and Vice-President
of the National Wildlife Conference. Finley's initial comment
was to the effect, "I think somebody's trying to beat the story of
the sea lion in the Willamette..." which referred to a previous
incident in which a marine mammal had undertaken a solitary
migration up a local Oregon river system. Finley followed this
with an even stronger position, commenting that the captured
octopus:

> "...couldn't possibly have traveled up the Co-
> lumbia (river) over the Cascades and the Celilo
> Falls, and up over the falls of the Deschutes and
> into the Metolius."

And concluding further:

> "If it is an octopus, and was in the Metolius
> River, somebody must have put it there, unless
> there is such as thing as a freshwater cephalopod,
> and if so, it's absolutely new to science."

Edward F. Averill, ex-game commissioner for Oregon, suggested that the octopus might not be the work of a local hoax.

Leaving on the 16th, Stanley Jewett, while in transit from Portland to Bend, Oregon, stopped off at Redmond to have a look at the octopus. On the basis of his examination, Jewett, who was later to achieve fame as the author of a leading book on Oregon birds, sent his report in a telegram to Portland's largest newspaper, *The Oregonian*: "The Redmond octopus is real, but I am not convinced of its origin..."

Thus, there was a tangible carcass of an octopus, but the question remained as to how it actually managed to arrive in the middle of Oregon, several hundred miles from the Pacific Ocean. After all, none of the wildlife experts would even consider endorsing the idea that a marine octopus could make its way unaided up the several hundred miles of the Columbia River and its tributary the Deschutes River to arrive in the Metolius River, where it was captured. After all, sea creatures are not normally able to survive several hundred miles of freshwater streams. Furthermore, none of the experts were willing to consider the possibility that a freshwater species of octopus existed; after all, none had ever been captured and identified before. (31)

A wryly and skeptical criticism was published in the same-day edition of *The Oregonian* (32) in the editorial column. Though stating that some marine denizens had adapted to freshwater environments, such as the freshwater sharks of Lake Nicaragua in Central America, "These are recognized repudiations of common habit—but an octopus in a trout stream, sir, that is another matter entirely." The column served to heighten doubt regarding the authenticity of the "freshwater octopus." The article further alluded that the now famous capturer of the Metolius

River octopus should be placed in the same catalogue as others who had seen the most infamous of all the sea's unknown creatures:

> "Mr. McDaniels of Redmond has outserpented the sea-serpent. If he has what he vows he has, he shall stand before kings, but their majesties, with the merest smattering of biological knowledge won't believe it, either."

These words proved prophetic and also absolutely correct. On the following day, the 17th, the postmortem results were revealed. George Aitken, of the Deschutes County Sportsmen's Association revealed that:

> "The statement that the octopus was taken from the Metolius River by McDaniels is entirely correct."

He then added a postmortem punchline:

> "However, the octopus arrived here overland from California about 10 days ago, I have been reliably informed." (34)

W.S. Rice, secretary of the Oregon Wildlife Council, confirmed the report. For the record, he showed that the octopus had been preserved in brine and was "very dead" long before McDaniels had fished it out of a deep pool. (35)

Thus, the true origin of the Metolius River octopus was established as being nothing more than a California marine resident that had been transported inland and ultimately dumped into the freshwater river. By all appearances and accounts, Mr. McDaniels, who had originally fished the octopus out of the river was an innocent victim of circumstances, an innocent bystander. Again, this case of an out-of-place cephalopod must reluctantly be relegated to that of a local hoax.

The Kankakke, Illinois, Squid

Later that year, another report of an out-of-place cephalopod came from the center of the nation, near Kankakke, Illinois. On the night of June 29th, Scotty Waldron, a janitor at the Domestic Laundry in town, was attracted to the Grinnell Ditch by his cat—named Tenoume—which was batting something on the bank of the stream. The cat's interest was in a squid, very much alive and with its tentacles wrapped partly around the head of the cat. The struggle between cat and squid continued for several minutes before the cat was able to dispatch it with a bite to the head. The squid was pale yellow, almost white in appearance and about a foot in length with ten tentacles. Waldron took the squid to City Hall where Charles Gregg took possession and preserved it in a jar of alcohol. The following day the preserved squid was placed on exhibit at Koehler's downtown drug store. (36, 37) It took several more days before news of this squid made the Associated Press wireservice. (38)

Although confirmation exists of the reality of the squid and also of Scott Waldron (37) as a resident of the city, what are we to make of his living squid? Should we conjecture that all of the excitement is over Waldron's cat attacking discarded fish bait? This seems the most likely explanation, but the question still arises as to how this squid was transported 1,000 miles inland to make its appearance in the ditch. It had to be transported overland by railroad, as the era of interstate trucking was still decades into the future. But why should someone toss out good fish bait in those tough economic times? Therein lies the real mystery! (39)

Take Me Back to Old West Virginia

A decade would pass before new accounts of out-of-place cephalopods in freshwater environments would again appear in the news. On January 10th, 1946, a 12-year-old student, James Largent brought a two and a half foot octopus that had been found in the Blackwater River near the Thomas railroad station to his school. A second octopus was brought to the school by brothers Philip and James LaMont and their friend, Kenneth Liller. Both had been shot with a small rifle. Yet a third octopus, about 15 pounds in weight, was shot but then dumped back into the river as the young students were unaware of the uniqueness of these cephalopds in freshwater streams. A few days later, still another octopus was shot in an impounded lake created by the Fairfax Electric Company power dam. A sixth octopus was found some distance below the dam and was brought to the attention of Principal Stelman W. Harper of Thomas High School. Dr. S. Benton Talbott, of the Davis and Elkins College, said that he was unable to explain how so many marine creatures had reached the Blackwater River. (40)

Two photographs were published the following day, the 17th, on the front page of the local newspaper, the Morgantown *Post*. One showed students Liller and Largent with their .22 caliber rifle at the north fork of the Blackwater River near the spot where they had killed the five octopuses. The other showed Principal Harper scratching his head while staring down at an 8-tentacled octopus as science teacher Miss Mary Colabrese examined it. (41, 42, 43)

Professor L. H. Taylor, zoologist at the local university, confirmed that the animals were indeed octopuses, but he was understandably unable to explain how they got into the local river located many hundreds of miles from the nearest marine habitat. He hoped to obtain one for further study. (44)

The following day, the 13th, Professor Taylor was skeptical regarding student claims that the octopus had been caught alive; he proposed that the swift, turbulent current in the river might have caused the agitation of the already dead carcasses so as to appear that they were in fact alive when caught and/or shot. (44)

An Italian storekeeper informed him that the octopuses may have been discarded by someone who had received them in a shipment from a warmer climate. Octopus is considered a delicacy by many Italians and they may have been shipped in ice in a manner to be properly preserved when they reached the town of Thomas in Tucker County. This claim was supported by the fact that the first of the creatures were noticed in the river near the local railroad station. Someone may have simply tossed these unwanted food delicacies into the river to dispose of them quickly.

With the passage of another day, on the 14th, a sixth octopus was captured by Thomas High School senior, Teddy Peters, age 15, in the same region of the river as the other five were taken. He stated that the animal was alive and holding onto a rock with its tentacles in a shallow part of the stream. This sixth octopus weighed 20 pounds and measured 40 inches from tip to tip. (45)

The local express agent confirmed that there had been no shipments containing octopus received. He had thrown away some salt fish which were received spoiled, but there was no octopuses. (46)

The Tucker County octopus controversy finally ended on the 19th, when a very tongue-in-cheek resolution was offered by Dwight Peter, secretary of the West Virginia Sportsmen's Association at a meeting whereby in part read: (47)

> "Now therefore, be it further resolved that we respectfully petition the West Virginia conservation commission, through its managing director here and now present, that a closed season be established on octopi in Tucker County; that every step be taken to insure their conservation and propagation with the distinct understanding that no adverse legislation be enacted affecting the present open season in Tucker County on certain species of hammerhead sharks, sulphur bottom whales, Lynn Haven oysters, Alaskan seals and lobsters *a la* Newburg."

The most likely explanation is that these six West Virginia octopuses were an unwanted or spoiled shipment for food, but were tossed away into the nearby river. And as such, should be listed as so.

For whatever reason, the state of West Virginia seems to have an affinity for reports of freshwater octopuses appearing in freshwater rivers. A decade later, a circa 1954 report originally published in the Pittsburg *Press*, was subsequently condensed and reprinted as:

> "An octopus in a creek near Grafton, West Vir-
> ginia. Found by four nameless boys who gave it to
> a dog catcher. It had two foot long tentacles. It
> died shortly after capture."

Additional details about this capture were not recorded, but the town of Grafton is east of Clarksburg, about half-way to Tucker County. (48, 49)

Another decade passed. In the Questions and Answers column of the spring 1965 issue of *Sea Secrets* a writer by the initials of C. T., of Hyde Park, New York asked: "While at Mohonk Lake in the Catskills I heard people talking about a freshwater octopus in the lake. Are there freshwater octopuses?"

To this question the editor replied, "There are no cephalopods, either octopuses or squids, who can survive in freshwater." (50)

Inquiries to the New York State Department of Environmental Conservation revealed that Mohonk Lake is a private lake 15 acres across in surface area with a maximum depth of 65 feet. It is located in the village of Mohonk , town of New Paltz, about 4 miles west of the city center and about 2 miles southwest of the Mohonk Preserve. Water quality of the lake is good and is intended for potable water intake and for trout. An 1987 Adirondack lake survey revealed 8 different fish species—but no cephalopods. (51)

This 1965 inquiry should therefore be relegated as a misidentification?

Last 20th Century Report

With the close of the 20th Century, the last documented account of the discovery of an out-of-place octopus was reported
in the last year from the banks of the Ohio River on November
21st, 1999. The octopus was discovered on the fossil beds of the
Falls of the Ohio State Park. Investigator Brad LaGrange was
able to acquire a photograph taken of this beached octopus by
Paul McLean, who worked for the nature interpretative center
of the park. The photograph later proved valuable in solving
the mystery of the misplaced octopus. McLean provided the following rather meager facts regarding the discovery: "It was not
alive and it was not in a state of decomposition. It probably
weighed less than a pound and the color shown in the picture is
close to accurate." Dominic Foster, who maintained a freshwater and marine center aquaria, identified the specimen as an
Atlantic octopus. A special exhibit was created for the octopus
when it was first discovered, but it was later discontinued.

From the National Resource Center for Cephalopods, research scientist John W. Forsythe, upon examining the photograph, proposed that the species was really *Octopus burryi*, an
octopus of the eastern seaboard that is also found in the Gulf of
Mexico and throughout the Caribbean. It is also possible that
the octopus was *Octopus filosus*, which has a similar body morphology. Both octopod species are readily available for the exotic
pet lover in the aquaria trade from Haiti.

It is certainly possible that someone had tired of keeping a
large and probably still growing exotic pet octopus and decided
to toss it into the local waters. An even more intriguing possibility is that the octopus had hitched itself a ride on a vessel
inbound from the Gulf of Mexico and somehow survived long
enough to reach the state of Ohio via the Mississippi River and
then the Ohio River transit. This, admittedly, is a highly remote
possibility, but given the number of barges that ply the river
trade this is not quite as outlandish as it might initially seem.
After all, we can name a number of species that have hitched
rides to launch their movements from one continent to another,

beginning with the Norway Rat (*Rattus norwegicus*) and House Mouse (*Mus musculis*) and ending with the Zebra Mussel.

New Reports in the 21st Century

Reports of live captures of freshwater cephalopods are still being recorded in the 21st century, at least in the southern state of Arkansas.

While visiting Conway, Arkansas, John Mazurek, Sr., from Glen Ellyn, Illinois, decided to go fishing in Lake Conway. Lake Conway is in the geographic center of the state. It drains into a small stream that in turn empties into the Arkansas River.

On the Monday of December 8, 2003, John Mazurek spotted an octopus clinging to one of the gates at the lake's dam and grabbed it. Mazurek's good-sized freshwater prize was sold to Arkansas Game and Fish Commission wildlife officer, John Harpher.

The origin of this 3-foot octopus in Lake Conway is unknown. The lake is connected via a tributary to the Arkansas River, but we are not seriously suggesting that the octopus made its way up the Mississippi River to the Arkansas River and then into Lake Conway all on its own.

It is much more probably that a local aquarium owner dumped it into the lake shortly before Mazurek spotted it and was able to capture it alive. (52)

Discussion and Conclusions

Unfortunately, we cannot conclude or even suggest that any of the reported and recorded instances of out-of-place freshwater cephalopods represent invasive species spreading into freshwater environments from marine waters. We also have to add that there is no information or data that allows us to suggest that these reported cephalopods represent a new and previously undiscovered squid or octopus able to survive in freshwater systems.

More likely, they collectively present a mixture of hoaxes, lost or discarded fish bait, discarded seafood shipped to inland markets, or marine aquaria pets that were discarded or deliberately released into freshwater streams.

The possibility of freshwater octopuses and squids remains a distinct but distant possibility given the current mobility of the human population. We suggest that the first valid reports of future freshwater cephalopods will occur immediately adjacent to well traveled near shore lanes such as inland waterway, canals, or somewhere along tidal rivers.

Chronological List of Out-of-Place Freshwater Cephalopods.

Date	Freshwater System	Cephalopod	Source/Fate
1902	Onondaga Lake, NY	Squid	Discarded fish bait?
1933	Kanawha River, WV	Octopus	Hoax?
1936	Metolius River, OR	Octopus	Discarded food/bait?
1936	Kankakke, IL	Squid	Discarded food?
1946	Thomas, WV	Octopuses	Discarded food/bait?
1954	Creek near Grafton, WV	Unidentified	Unknown
1965	Mohonk Lake, NY	Octopus	Mistaken identity?
1999	Ohio River, OH	Octopus	Aquaria pet?
2003	Lake Conway, AR	Octopus	Aquaria pet?

Acknowledgements

We thank the following individuals for their help in supplying materials and references for this chapter: Chad Arment of Landisville, Pennsylvania, for copies of the *North American Biofortean Review*, of which he is editor, for help in references on this topic. Scott Mardis of Winooski, Vermont, for science references of rumors of freshwater cephalopods. Mark A. Hall of Wilmington, North Carolina, editor and publisher of *Wonders*, for copies of various West Virginia newspapers that carried accounts of "freshwater octopuses" captured in that state. Susan H. Harper, West Virginia Historian and Archivist, Kanawha Co. Public Library, Charleston, West Virginia. Scott A. Kishbaugh, Environmental Engineer, Bureau of Watershed Management, Albany, New York. Alexsander Lovcanski provided sketches of squid and octopus.

Literature Cited:

1) Anonymous. Here's that strange Florida monster. *Rocky Mountain Husbandman* (Great Falls, Montana). 10 October 1935.
2) Anon. Lizards at large roaming Florida. Connecticut *Post*. 23 June 1970.
3) Anon. Great lizards invade Florida. *National Examiner*. 1981. Number 122.
4) Anon. Jumbo Jumpier. New Haven *Register*. 2 August 1985.
5) Anon. Poisonous giant toads spread in Florida. New Haven *Register*.
6) Coleman, Loren. 1979. Alligators-in-the-sewers. A journalistic origin. *Journal of American Folklore*. 92: 335-338.
7) Anon. 4-foot alligator is captured in East River. New York *Herald Tribune*. 1 June 1937.

8) McFeathers, Dale. Remember "The Birds" around you. New Haven *Register*. 2 July 2001.

9) Singleton, Don. Giant reptiles in sewers. New York *Daily News*. 5 June 1988.

10) Anon. An alligator captured. Hartford *Courant*. 10 July 1902.

11) Bernstein, Michael. "Alligator" in West Hartford. Hartford *Courant*. 1 May 1966.

12) Anon. Fisherman nets crocodile in lake. Connecticut *Post*. 30 July 1996.

13) Anon. Fishermen nets 3-foot crocodile. New Haven *Register*. 30 July 1996.

14) Fredricksen, Lynn. A gator in Guilford. Or is it just a croc? New Haven *Register*. 9 July 1999.

15) Fredricksen, Lynn. Gator watch will continue, just in case. New Haven *Register*. 10 July 1999.

16) Anon. Goose-gobbling "gator" sighted. Connecticut *Post*. 10 July 1999.

17) Anon. Alligator shot in southern Maine. Kennebec *Journal*. 20 September 2000.

18) Anon. Habitat: Nova Scotia's mysterious interloper. *Equinox*. January-February 1987.

19) Arment, Chad, and Brad LaGrange. A freshwater octopus? *North American Biofortean Review*. December 2000. Volume 2. Issue 5.

20) Kershaw, Diana. 1984. *Animal Diversity*. University Tutorial Press. Wiltshire, Great Britain.

21) Rupert, Edward E., and Robert D. Barnes. 1994. *Invertebrate Zoology*. 6th Edition. Saunders College Publishing Company. New York.

22) Clark, John M. 1902. The squids from Onondaga Lake, New York. *Science* 12: 947-948.

23) Ortmann, A. E. 1903. *Illex illecebrosus* (Lesueur), the squid from Onondaga Lake, New York. *Science* 17: 30-31.

24) C.T. Question Column. 1965. *Sea Secrets*. December 1965.

25) Anon. Jokes on scientists. Washington *Post*. 22 February 1903.

26) Anon. Get octopus in Kanawha, New York. New York *Times*. December 1933.

27) Anon. Octopus caught by two boatman on Kanawha River. Charleston *Gazette*. 25 December 1933.

28) Anon. Devil fish may join capitol's archives. Charleston *Gazette*. 27 December 1933.

29) Anon. Octopus story is just a hoax. Charleston *Gazette*. 29 December 1933.

30) Anon. Octopus reported captured in central Oregon river. New York *Herald Tribune*. 16 March 1936.

31) Anon. Octopus report stirs skepticism. *The Oregonian*. 16 March 1936.

32) Anon. Questions octopus tale. New York *Times*. 17 March 1936.

33) Anon. Redmond octopus declared genuine. *The Oregonian*. 17 March 1936.

34) Anon. From the limpid Metolius. *The Oregonian*. 17 March 1936.

35) Anon. Redmond octopus from California. *The Oregonian*. 18 March 1936.

36) Anon. Sea monster captured in Grinnell Ditch, octopus-like creature and cat in battle. Kankakee *Journal*. 30 June 1936.

37) Anon. Sea monster is identified as a squid, octopus' cousin. Kankakee *Journal*. 1 July 1936.

38) Anon. Cat kills stray octopus. *Herald Tribune*. 4 July 1936.

39) Polk, R. L. 1937. *Polk's Kankakee (Kankakee County, Ill) City Directory. 1937. Including Aroma Park, Bonfield, Bardley, Bourbonnais, Buckingham, Essex, Grant Park, Herscher, Irwin, Manteno, Momence, St. Anne, and Wichert.* Chicago, Illinois.

40) Anon. Schoolboys find five octopuses in state stream. Morgantown *Daily Mail*. 11 July 1946.

41) Anon. Eleven octopuses found near Thomas. Morgantown *Post*. 11 January 1946.

42) Anon. Biggest fish story of the year puzzles Thomas teachers.—mystery of octopuses is unsolved. Morgantown *Post*. 12 January 1946.

43) Anon. Octopuses found in state real, puzzled West Virginia University zoologist admits. Charleston *Daily Mail*. 13 January 1946.

44) Anon. Professor doubts octopus living. Charleston *Gazette*. 14 January 1946.

45) Anon. Youth catches sixth octopus with string. Morgantown *Post*. 14 January 1946.

46) Anon. Train agent spikes one octopus theory. Morgantown *Post*. 16 January 1946.

47) Anon. State sportsmen seeking "closed octopus season." Morgantown *Post*. 19 January 1946.

48) Anon. 1955. Displaced critters. *Doubt*. Volume II.

49) Hall, Mark A. 2001. Mysteries of West Virginia. *Wonders* 6: 113-126.

50) Kishbaugh, Scott A. Private communication. August 2002.

51) Anon. 1987. *Limnology studies of Mohonk Lake*. Adirondack Lake Survey Cooperation. New York.

52) Mosby, Joe. Illinois fisherman "hooks" octopus at Lake Conway. Arkansas News Bureau. 14 December 2003.

Preliminary Notes on a North American "Flying Snake"

Chad Arment

Does a Flying Snake exist? Despite reports to the contrary, scientists insist that no known species of snake actually flies. But a cousin of mine declares he has seen a flying snake. He claims one attacked him in New Rochelle, New York, and chased him up a lamp-post while he was on his way to join the local chapter of Alcoholics Anonymous.

—Henny Youngman

Mythologies on several continents refer to flying serpents, from the Mesoamerican Quetzalcoatl to Egyptian representations of winged cobras. Most of these are imaginative renderings combining divine or supernatural attributes (symbolically represented as wings) with a creature that, depending upon culture, is treated with fear or deference. A few sightings of "winged snakes" have been discussed in cryptozoological literature. Sometimes these are labeled "living pterosaurs," disregarding details that don't fit nicely within that classification. Rarely, the supposed animal is noted as having a truly serpentine form. See, for example, Dr. Karl P. N. Shuker's discussion of certain Welsh folklore in *From Flying Toads to Snakes With Wings* (1997, St. Paul, MN: Llewellyn).

In North America, focused investigation suggests that a "flying snake" is a legitimate cryptid, reported both in the past and
in the present. This cryptid appears to have been downplayed
within cryptozoology for several reasons: first, the historical
reports are sporadic, and not easily found through random
newspaper searches; second, the apparent anatomical absurdity
of such a creature makes it easy to dismiss without due consideration; third, most modern sightings are reported from Native
American reservation lands, particularly in southwestern states,
where "flying snakes" may be part of the traditional ethnofauna,
but sacred status may make it difficult for outsiders to learn
about them.

A few years ago, a small loose-knit group of six or seven
cryptozoology researchers interested in oddball mystery animals
formed an email discussion list. This allows us to toss around
ideas for research and speculation, a process that is difficult to
accomplish in larger group settings where diverse personalities
hold hard-nosed opinions. Through cooperative solicitation efforts
(advertisements and letters to the editor in rural newspapers),
sharing reports that come in through websites, interviewing
witnesses, examining field sites, historical research, and other
activities, we have managed to locate and evaluate a number of
intriguing accounts. This chapter, along with Nick Sucik's chapter
on the "dinosaur" reports, introduces a few of the cryptids with
which we are particularly interested.

What we are presenting here is an initial description of a
cryptid, but not, of course, confirmative evidence. There may
not be a biological explanation for this mystery animal—it may
be nothing more than a string of coincidental anomalies resulting from poor communication, misdesciption of known species,
or poetic license on a slow news day. But, cryptozoology is first
and foremost an investigative methodology—and there are
plenty of starting points for investigation in this case.

The first glimpse of this animal is found in Charles Fort's
classic text, *Lo!*, where he briefly notes three reports of large
serpentine "creatures" being seeing in the air. One is clearly

meteorological phenomenon (a snake-like formation encircling the disk of the sun as it rose, *New York Times*, July 7, 1873), while another has no easily-recognized biological explanation ("... something resembling an enormous serpent floating in a cloud ... as large and long as a telegraph pole, was of a yellow striped color, and seemed to float along without any effort. They could see it coil itself up, turn over, and thrust forward its huge head as if striking at something," *New York Times*, July 6, 1873.) But the third involves a sighting in Darlington Co., South Carolina, that clearly suggests a living animal. The news brief that Fort cited from the *New York Times*, May 30, 1888, is:

A Flying Serpent.

Columbia, S.C., May 29. — Closely following the appearance of the hand of flame in the heavens above Ohio comes a story from Darlington County, in this State, of a flying serpent. Last Sunday evening, just before sunset, Miss Ida Davis and her two younger sisters were strolling through the woods, when they were suddenly startled by the appearance of a huge serpent moving through the air above them. The serpent was distant only two or three rods when they first beheld it, and was sailing through the air with a speed equal to that of a hawk or buzzard, but without any visible means of propulsion. Its movements in its flight resembled those of a snake, and it looked a formidable object as it wound its way along, being apparently about 15 feet in length. The girls stood amazed and followed it with their eyes until it was lost to view in the distance. The flying serpent was also seen by a number of people in other parts of the county early in the afternoon of the same day, and by those it is represented as emitting a hissing noise which could be distinctly heard. The negroes in that section are greatly excited over the matter. Religious revival meetings have been inaugurated in all their churches, and many of them declare that the day of judgment is near at hand.

An additional reprint of this news item from another newspaper (Coshocton, Ohio, *Semi Weekly Age*, June 1, 1888) adds that the animal was "as large around as a good-sized human thigh." It also states that the incident happened on Monday evening rather than Sunday evening. The question of timing is not determinable, unfortunately, without seeing the original South Carolina newspaper story. Locating the original story, and searching the same paper for the months prior and after this date, might lead to other sightings in the same area.

With some digging into a digital newspaper archive, two more early "flying snake" reports were found. The first comes from Leavenworth, Kansas, reprinted in an Ohio newspaper:

A Flying Snake

A few weeks ago we referred to a lady living in the southern part of the city having seen a flying snake in her peregrinations thought [sic] that delightful portion of the metropolis. At the same time we published the statement of an aged woman, a soothsayer, who predicted that in a short time the air would be full of flying serpents. Yesterday we were met by a friend, who inquired, in an excited manner, if we had ever seen a snake that had wings, and "flew through the air with the greatest of ease?" From his statement we learn that while two boys named Remington and Jenkins, the former from this city, and the latter a Platte countian, were hunting in the woods, a serpent was seen approaching them, about four feet above the earth. Jenkins took off his hat and throwing it over the snake, succeeded in capturing it. It is over one foot long, spotted, and has wings about the size of a man's hand. The boys have the serpent preserved in alcohol.— [Leavenworth *Times*.]

—Athens (OH) *Messenger*, Sept. 16, 1875.

This, of course, is opportunity for an investigator from eastern Kansas or neighboring Missouri to determine what happened to this intriguing specimen, or whether the Leavenworth *Times* published further details (perhaps an illustration) of the animal.

The second newspaper account involves a report from Taylor County, Iowa:

The Bedford *Times-Independent* of last week tells a remarkable story about a flying serpent, claimed to have been seen by a gentleman named Corder living near that place. Mr. Corder says when first seen it resembled a buzzard, but as it drew nearer its appearance was different from any flying animal he had ever seen. As it descended lower and its outline became more distinct, it took the form of a great serpent, writhing and twisting, with protruding eyes and forked tongue. Great scales, which glistened in the sunlight, covered its huge body, which appeared to be flat and nearly a foot in width. While they were gazing at it with awe and astonishment, it landed in a cornfield, a few rods distant, with a dull thud. Those who saw it were so frightened that they did not dare to go to the field in search of it, and it was allowed to pass on its way unmolested. Many theories have been advocated to us as to the probable nativity of the insect but none appear plausible. The Times also adds: "Perhaps our readers will be inclined to doubt the truth of this story, but if any such will call on Lee Corder, or any of the family, who reside about 5 miles from this city, they may be convinced of its truth, as they are people of unimpeachable veracity."

—Humeston (Iowa) *New Era*, Aug. 11, 1887.

These newspaper reports are primarily nineteenth-century, but one is found in the early 1900s. A very brief account was published in the Decatur, Illinois, *Review* (October 14, 1917) regarding a Florida incident.

Snake Turns Aviator

Kissimmee, Fla. Jim Sanitu, Seminole indian from the Everglades, came to town today with the skeleton of a winged snake, which he says he killed after a hard battle. Prof. Hedges, principal of the local high school, sent him back after the skin.

Not much to go on, but British researcher Richard Muirhead has noted that Goldsmid's (1886) *Un-Natural History, or Myths of Ancient Science* quotes early explorer Hieronymu Benzo, from the French exploration of Florida: "I saw a certain kind of Serpent which was furnished with wings, and which was killed near a wood by some of our men. Its wings were so shaped that by moving them it could raise itself from the ground and fly along, but only at a very short distance from the earth."

Nick Sucik recently had opportunity to attend a university lecture by writer Adrienne Mayor. Mayor has suggested that some mythical creatures from Greek, Roman, Native American, and other legends may be based on early interpretations of fossilized animals. Looking through her most recent book, *Fossil Legends of the First Americans* (2005, Princeton Univ. Press), Nick noticed that Mayor includes an account of a flying snake.

In this story, a Crow medicine man, Goes Ahead, went on a vision quest in the Wolf Mountains of Montana around 1870. After fasting for a period, Goes Ahead noticed what looked like a small bird flying awkwardly. As it approached him, he realized it looked reptilian, "serpentine," "lizardlike," with a long tail and "wings something like a dragonfly's." Goes Ahead picked the animal up, and incorporated its body into his medicine bundle. He kept the bundle for about thirty years, eventually throwing it away with other sacred possessions in 1900 upon baptism as a Christian. Mayor notes that after surviving the Battle of Little Big Horn in 1876, (going into battle even after he and other scouts warned Custer of their poor chances), Goes Ahead carved a representation of the winged snake on a tree in a pine ridge about three miles east of the battleground. This was relocated and photographed by historian William Boyes in 1973. While the photo has been lost, Boyes recalled a carving 15-18 inches long, a blunt snakelike body, and two pairs of dragonfly-like wings. Mayor's suggestion was that Goes Ahead had picked up a fossil of some sort, perhaps a large dragonfly or petrified bat, but notes that the famed Chief Medicine Crow also described a winged snake when sketching local reptiles.

Nick has gathered sighting accounts in Arizona, documenting and photographing folkloric claims such as large stick "nests" or "towers" supposedly created by these animals. Within the Navajo culture, these are referred to by a phrase that roughly translates to Serpentine-Animal-That-Flies. Within Hopi tradition, they are known as Sun Snakes. Other tribal groups refer to them as Lightning Snakes, while in Kiowa traditions, as told to Nick by Native American author Russell Bates, Thunderbird is associated with a large glittering flying snake, similar to the Cherokee's Uktena. Archaeological finds note the presence of winged serpents, as well. Caddo designs in Texas and pottery excavated in Moundville, Alabama, depict flying serpents. Thermal imagery of the Great Serpent Mound in Adams Co., Ohio, suggests that the original mound had winglike structures.

One of the more interesting, and earliest, New World stories of a "flying snake" is that of the Quetzelcoatl. An encyclopedia of Aztec society published in the 1500s, the *Florentine Codex*, contains an entry for this: "He was not only the god, or the King of Tollan [Tula], but also a particular, small venomous snake from ... the former region of the Olmec... Its toxicity and mode of attack were similar to the arboreal palm viper. 'In order to bite you' ... 'first it flies, quite high up, well up it goes, and it just descends, upon whom or what it bites. And when it flies or descends, a great wind blows. Wherever it goes, it flies.'"

Due to cultural sensitivities on the Native American lands on which present-day sightings occur, we are not yet able to present the more recent sightings. Nick's field research continues, however, and he has received encouraging support by many elders and officials among the people living in this amazing region.

At this point, it isn't feasible to propose a biological candidate for this cryptid. While some features appear reptilian, we can't rule out the possibility that a large invertebrate may be responsible. Generally speaking, the animal probably isn't too big, with exaggeration responsible for claims of larger sizes. The ethnoknown animal is not pterosaur-like, but has what may be "dragonfly-like" or transparent membranous wings, capable of active "fluttering." This transparency (particularly noted in a briefly captured "flying snake" report that Nick heard from a Navajo family) may be responsible for what is sometimes noted as apparent lack of means of propulsion. The body appears to contort in flight, even to the extent of "corkscrewing" in the air. (Interestingly, while this mystery animal has no direct relationship to the Asian gliding snakes, genus *Chrysopelea*, those snakes consciously modify their descent through undulation in the air.) The wing movement may also be responsible for a "hissing" sound noted in both historic and modern accounts.

If a true biological species is responsible for this mystery animal, it probably will be very different anatomically from other modern animals—particularly if it is a reptile. Still, it may not be so far from Reptilia proper. There are intriguing little fossil reptiles like *Icarosaurus* and *Longisquama* suggesting that unique modes of flight are quirks not unexpected in the tapestry of life.

The "flying snake" is a strange, yet certainly ethnoknown, mystery animal here in North America. Dr. Bernard Heuvelmans was correct when he pointed to cryptozoology as a fundamental investigative methodology—the path to zoological discovery begins in learning what we can from cultures already familiar with the animal. Yet we must move beyond ethnozoology toward a critical and scientific evaluation of confirmative evidence. Our goal for this introduction is to stimulate further research in hopes of obtaining such evidence.

Crypto New Mexico

Jerry A. Padilla

New Mexico, the Land of Enchantment, is one of the oldest multicultural settled regions in the continental United States. This is an ancient land where several Native American dialects and Spanish are still spoken, along with English, introduced in the early-mid Nineteenth Century. New Mexico's multicultural folklore is rich, and while many of these tales came about as a way to scare little children away from danger, perhaps some came about as a way to rationally explain a few creatures that didn't fit the norm. Spanish-speaking immigrants continue to come here looking for a better life, and along with new residents from other parts of the nation and the world, become enchanted and stay.

When they become permanent residents, newcomers also contribute to contemporary folklore, as in the case of the *chupacabra*, and exchange tales about *naguales*, (shape shifters), giant birds of prey, the onza, sasquatch, *loupe garu*, and various and sundry other legends and creatures. Some knew about crypto-creatures in their homelands or other states, and soon realize part of the enchantment that is this state of New Mexico is our unique folklore.

El Chupacabra, the goatsucker, is not just a Puerto Rican and south-of-the-border entity anymore. Not if recent sightings in Las Vegas, New Mexico, the other Las Vegas, are taken seriously.

Perhaps some folklore is based in part on cryptids. There is a line in *La Pastorela*, The Shepherd's Play performed during Christmas season, where the character of Michael the Archangel accosts the character of Lucifer, calling him "feo dragon basilisco, ugly basilisk dragon." Historian and columnist Marc Simmons recently made reference to the basilisk, the legend of a reptilian dragon-like creature against which the only defense was to turn it's own reflection on it with a mirror. Is the basilisk merely folklore, or did Spanish Colonials familiar with caimans, crocodiles, and alligators in Mexico encounter something similar enroute to Northern New Mexico along the southern reaches of the Rio Grande near El Paso? While familiar with the language, and morality plays, the term *basilisco* might have been used by the common people to describe something that reminded them of dragon legends. Could crocodilian creatures still have been found along the Río Grande which forms part of the current border with El Paso and much of Texas/Mexico during the 1500s? Extra large catfish have always been reported in this river and it's reservoirs. Folks seeing an alligator gar or its remains on river banks, for the first time, could have easily described this toothy primitive looking fish as a basilisk.

Giant catfish are reputed to live in the depths of big lakes at Navajo Dam and Elephant Butte. A popular modern legend relayed by G. Garner Jr., says no evidence was found of body parts during investigations of the David Parker Ray murders at Truth or Consequences, "because some say he was cutting up his victims and dropping them into Elephant Butte Lake at night so the really big catfish would dispose of them."

Then there are *los abuelos*, translated literally, *grandfathers*, or in the plural, *grandparents*. These scary entities are also part of New Mexican folklore, described by folklorists "as part human, part animal, who live in the mountains, guardians of our culture. To see an *abuelo* in summer is very bad luck, but they usually don't come out of their mountain habitat except in winter, when many of our traditions are being performed. They don't talk, except to ask children and wayward adults if they remember their prayers. If you see one, you know you haven't been behaving."

There are contemporary Hispanic New Mexicans who remember having encountered an *abuelo* when their parents had to tell them to behave, or not wander from home after dark. The sight of the dreaded *abuelo* was usually enough to encourage them not to wander away, or adjust their behavior and attitude in a hurry.

An *abuelo* was another of those creatures of folklore that had a way of appearing when children refused to behave, or when a potential witness was needed when adults were about to commit an injustice. As Indo-Hispano children got older, they came to know who and what *abuelos* were. Usually an older relative, most likely male, would don a disguise, a rawhide mask, and might look like a gorilla, a bear, or other creature, but oddly enough, walking like a human.

Did this tradition come from medieval Spain as some claim, or did traditions about mountain creatures, resembling both human and animal come about from encounters with *el Patas Grandes*, as Bigfoot is known in old Mexico? Is it possible New Mexicans have been witness to evidence of surviving populations of *Gigantopithecus* since the arrival of those first sixteenth century Spanish speaking colonists? And, blended eyewitness encounters with medieval folklore?

Did accounts about *Los Descalsos*, the *barefoot ones*, told by some old time shepherds and herders, come about when they encountered large tracks resembling a naked human foot? The pragmatic shepherd or mountaineer, upon seeing such tracks as others have reported in other parts of the continent, not feeling threatened, would have left it at that. Why spend too much time wondering about who or what less-fortunate being had made such tracks, when *lobo* wolves, coyotes, and mountain cats posed a real threat to sheep and cattle? They were simply made by the *barefoot ones*. Even if the tracks were oversized for a typical human, as the late Taos County rancher Pedro Lucero used to say about unexplained phenomena, "These mountains are so big, and there's so much rough country you can't get through even on horseback, there could be anything up there."

Bring up the subject of Sasquatch in contemporary times, among all walks of life, professions, levels of education, cultures

in New Mexico, and accounts about this elusive legendary creature are more common than sightings of mountain lions and
really big mule deer bucks, two well known indigenous species.
Most encounters happen at dusk, or just after sunset, sometimes
in remote alpine areas like the Pecos Wilderness, Indian reservations, less traveled parts of national forests, or in more recent accounts, out in the sagebrush deserts, prairies, and nearer
places not heavily populated. The commonality is of a shy, nocturnal, dark to reddish hair- or fur-covered creature, at least
eight feet tall, standing or walking upright, whose head slopes
from a low forehead to a pointy crown. Folks, almost all request-
ing to remain anonymous, agree it resembles a gorilla, but walks
or stands like a human.

Said R. Jackson, from northern Taos County, "We were near
where the Rio Grande Gorge starts, right near the Colorado line
on the north side of Ute Mountain, looking for interesting
shapes of dead wood, in an area that is as beautiful as it is remote.

"Other people claim they've encountered Bigfoot out there,"
he continued. "It's really wild and inaccessible, and as the sun
was starting to get low in the horizon, we could sense something watching us. As daylight started getting dimmer, we heard
the noise these creatures make, and right then we knew, it was
time to get out of here."

Concluding, Jackson said, "Whatever it is, it's not hurting
anyone, people should leave it alone, it's part of nature."

Oversized Owls and Birds

New Mexican folklore about owls usually involves tales about
how these nocturnal hunters are utilized by those practicing
witchcraft to disguise themselves. The premise of the *nagual*
as known in Mexico, or *shape shifter*, is the animal form taken
by those who with the knowledge to do so disguise themselves.
Stories tell of individuals with the power to temporarily take
the form of an animal, or bird, in order to keep their identity

secret for whatever purpose, in order to work evil on others without being found out, or in order to travel great distances.

The latter reason is given as an explanation because an animal can cover greater distances easier than a human. A winged creature such as an owl would not only afford these shape shifting individuals the ability to fly, but also to see well at night.

The late Elena Bustos Lucero of San Francisco, Colorado, and Taos, N. M., had an encounter in the late 1950s or early 1960s with an oversized owl in a rural area. Dusk approaching, she was putting tools away for the night coming in from working on her vegetable garden. A bird described by eyewitness Rosa M. Lucero, her grand-daughter, as "an owl with feathers on it's head that looked like horns, at least four to four and half feet tall, waddled out of a stand of red willows near the *acequia* (irrigation ditch) bordering the garden."

Lucero continued, "When my grandmother first saw this enormous owl, she stood her ground as it approached her. She asked me to bring a good-sized piece of wood from the woodpile. I was scared but since she showed no fear, I grabbed a piece of firewood, and took it to her. The owl would waddle up to within a few feet of my Granny, then turn and walk back towards the willows.

"She told me to bring her a bigger piece, one she could club the big owl with if necessary. I ran back with a bigger piece of firewood, with which Granny Lucero approached the owl, making the sign of the cross, brandished the club at the owl, threatening to hit it if it should come any nearer. It continued to walk back and forth in front of her, and as it got darker, walked back into the willows. When it didn't show itself anymore, my Grandmother took me by the hand, told me not to be afraid, and we went inside for the night. I had never seen before, nor since, seen such a large owl."

Lucero explained the owl never flew, but "sashayed back and forth near my Granny, as if challenging her. We only saw it that one time." Lucero grew up in rural Colorado and was familiar with owls, and other local wildlife, and affirms it looked like a great horned owl, didn't fly, didn't hoot or make any noise. "It just walked around in the garden by the willows. My grand-mother was convinced it was a *nagual*, someone taking the form of an owl, because she herself said that in all her long life she'd never encountered an owl so large and unafraid of people."

Other accounts of very large birds include stories from the Cimarron area about something referred to in local accounts as Baa Hawk. Interestingly enough, there is Comanche folklore about the *baa*, a very large bird of prey. Northeastern New Mexico where Cimarron is located was frequented by different bands of Comanche and their linguistic cousins the Utes in the 18th and 19th centuries. Comanche folklore tells of the great cannibal owl, a giant bird of prey, a dangerous threat taken serious by even the most valiant of hunter-warriors. It is a good chance the name *baa hawk* among Mexican-American settlers and their descendents came from interaction with Comanche and Utes. Similar accounts tell of a very large bird of prey that has been reported in the mountains, part of the Sangre de Cristo Range. Cimarron sits at the base of the mountains near the entrance to a canyon through which a river of the same name flows into the prairie. Headquarters of the well known Philmont Scout Ranch is a few miles south at Rayado. The prairies are immense, seemingly endless, and the mountains and mesas are rough, and somewhat inaccessible away from trails and ranch roads. The description of this giant bird of prey resemble that of the pre-historic roc. Before it became popular to blame unexplained cattle mutilations on UFOs or government experiments, these were attributed to the *baa hawk*. Sightings are usually near sunset or right after sunrise, alluding to nocturnal habits.

Robert Magill, journalist and a former Eagle Scout, familiar with the Philmont area, knowing all the peaks, mesas, trails, native flora and fauna, told of an encounter not too far from Cimarron. He related a brief yet very surprising account while camped at Charette Lake during June, 2005, near the border between Colfax and Mora Counties. Stirring from his sleeping bag at dawn to prepare breakfast he exclaimed "Damn, did you guys see that? That bird that just swooped overhead had at least a 17-foot wingspan! I heard it before I saw it." Magill explained to his partners it was flying so fast, it went completely out of sight towards a deep canyon across the lake.

Out-of-Place or Alien Animals

Oryx or spike bull elk, a case of mistaken identity?

Oryx or gemsbok are not strangers to New Mexico. Introduced on the White Sands Missile Range in the 1950s, oryx have found New Mexico quite to their liking and adaptability. So much so that they have spread beyond the confines of the military reservation, especially west where they have been reported running across I-25 between the towns of Socorro and Truth or Consequences. However, since the mid 1980s, there have been reports of these exotic African antelope in far northern New Mexico. These have been reported in the sage flats bordering the mountains of western Taos County, especially near the tiny community of Tres Piedras. It is not inconceivable that some may have, over the decades, wandered up the canyons of the Rio Grande, which bisects New Mexico north to south, and established themselves in this particular area. Deb Williams, who has lived on a ranch just south of Tres Piedras a few years, has heard about the oryx.

While they don't seem to be seen very often, there have been enough random sightings, that oryx have become a kind of local legend. Were they released from a defunct exotic game ranch reputed to have been a few miles east during the later 20th century? That's another possibility. M. Sullivan, native New Mexican, formerly of Miami, N.M. grew up on a ranch, is very familiar with all New Mexican wildlife. Sullivan, who has hunted oryx, says that for those who are not familiar with these beautiful creatures, they can be mistaken for a spike bull American elk. "Oryx run just like mules, but they are about the same size as bull elk, mules, and horses, and when you get up close you can tell the difference by the shape of their tail, their color and markings on their face... Some spike bull elk (a name for young one- or two-year-old bull elk) will grow extremely long single tined antlers the first year they grow antlers. It's possible people could mistake the two species," he explained.

"I once saw a bull elk taken on the Vermejo ranch that had only one antler that stuck straight out from it's forehead, like what people describe as a unicorn. It was a non-typical and stranger still, it had ring-like formations along the length of the single antler. As far as wild oryx roaming out of southern New Mexico, someone who knows what they really look like or that could get a photograph of them could answer that mystery."

Black Panthers

Images of black panthers are so popular among Hispanic New Mexicans, they can almost be called an icon. Many art forms, curios, and souvenir items representative of these wild black felines have been brought home from visits to Mexican border cities for years and are common in New Mexican households. The black panther is a popular image and it stands to reason, the black variety of jaguar being well known in Latin America. The melanistic jaguar, while known to exist in the wilds of Latin America from Mexico to South America, seems to be making appearances north of the border.

Myriam Morales, a health care professional, related an account of having seen what she describes as a "black panther" in Santa Fe, New Mexico's capital city. "A friend and I were exploring the outside of some old buildings in Santa Fe a few years ago (mid 1990s). It was getting dark, and we were at the old Montessori School on the road behind the old High Road to Taos. This is the vicinity of Calle Largo and Barranca St., where I saw the panther."

Morales continued, "At first I couldn't believe my eyes, and as I continued to watch the big cat as it moved silently and cautious, I knew that it couldn't anything but a large cat because of the way it moved. I know what I saw, a black panther. It wasn't visible for long, but long enough to ascertain it really was a very large cat, completely black, something you'd expect to see in a zoo, or out in the wild, not lurking in one of the oldest parts of Santa Fe."

Another account came second hand from a professional associate. "Some of my friends in Sunshine Valley (about 25 miles north of Taos) shared they and others have seen what they believe to be a jaguar in the rural area in which they live. Others have described a black panther. They said they think that animals native to Mexico and places farther south are gradually migrating north into areas where they are not bothered by humans, finding refuge in nearby national forests and places that are not heavily populated. That's how they explain reports of big cats that are not supposed to be in this area." (Name withheld by request, on file.)

Some people in Taos are of the opinion that our changing climate, habitat loss, and pressure from humans is the reason cats native to Mexico and Central America like the jaguar and jaguarundi have been reported in remote parts of New Mexico. And big cats people have tried keeping as pets, when it doesn't work out, are simply released into the wild. They adapt, find prey species, and in the case of the mystery creature in Las Vegas, N. M., feed on domestic dogs and cats.

Then there is the account about a maned lion roaming mountains and canyons northeast of Taos, also in the late 1980s. A couple of local Hispanic brothers where hunting deer, or elk when they encountered what they described as an African lion. They are from a family famous for their hunting abilities. Growing up on a ranch at the base of the Sangre de Cristo Mountains, they have hunted all their lives and are familiar with all local predator species, including bear, puma, coyote, bobcat, and the occasional lynx.

"There's no law protecting African-type lions in this state, so my brothers tried to bag it, but it seemed to know the area it was in really well, and got away from them," shared a younger brother who requested anonymity. (Name on file).

He continued, "It's rare when they can't catch something they are hunting once they spot it." It's the only account of this kind of lion being seen in the wild in northern New Mexico. How it got there is anybody's guess. Most likely it was released from a small traveling circus suffering hard times, or by someone who tried keeping it as a pet. This could well be the explanation for alien big cats reported throughout North America. Or is it possible there are still remnant populations of prehistoric American lions from primitive times still holding on in remote hard-to-reach areas. In Pre-Columbian times, jaguars ranged throughout the American Southwest, and possibly even into the deep South.

The Las Vegas Daily *Optic* reported October 25, 2005, that residents reported a strange creature locals began to call *chupacabra*. It all started with reports about cats and small dogs being killed and eaten by the unknown predator. Then Sider Esquibel Jr., reported seeing "a large catlike creature, dark in color with a black tail, walking upright." Esquibel said he was familiar with *chupacabra* stories but this creature was not like anything he had ever seen before. An experienced hunter, Esquibel also related the creature was not like anything he'd ever encountered afield. He continued explaining that when he shot at the creature that was outside his residence near a barn, the creature jumped some 70 feet up on the roof of the barn to escape. Apparently it got away.

A couple of women on their way home near where Esquibel had his encounter, saw the creature in the headlights of their car. They described it "with red eyes," and when they tried to run it over, the creature easily lept out the way of their vehicle, escaping over a standard four-strand barbed wire fence.

The *Optic* illustrated the front page article with a photograph of a track of the animal. It resembles a cat track. Is this the *chupacabra*, or as many have come to believe, an escaped or released black jaguar. After the initial rash of sightings, the excitement died down. A State Policeman stationed at Las Vegas, who requested anonymity said, "We had lots of calls for several days from people all over the county reporting what they called the *chupacabra*. We checked into the complaints, as routine procedure, but it seems to have moved on after a few days. People heard noises or saw something they weren't sure of, they'd call to say it's the *chupacabra*. Nobody else claimed to have seen it up close like Sider and his neighbors did."

Many are convinced that what was killing and eating domestic cats and dogs, and bothering farm animals is in fact a black jaguar.

There have been no more reports about the *chupacabra* or any other cryptid out of Las Vegas since. A mother and son had reported previously what they thought was a very dark mountain lion lurking on the edge of a nearby pasture on an early August 2005 morning.

El Pichucuate

New Mexico is home to several varieties of rattlesnake, the only venomous viper native to this state. Prairie rattler, western diamondback, and sidewinder are familiar to those interested in local reptiles. Folklore about serpents includes a story about a giant rattlesnake that lived in a kiva at the abandoned Pecos Pueblo, between Las Vegas and Santa Fe. According to one legend, the snake was so big and heavy that when it escaped confinement, people in the area were able to follow it's path because of the indentation it made on the ground over which it passed. It's trail led to the Rio Grande, into which it is reputed to have escaped forever.

Snakes are held in high esteem by many Native American cultures. Some old time Hispanic New Mexicans would not harm snakes, especially bull snakes, because they helped keep mice and rats in check on farms and ranches.

In Taos, the legend of El Viboron, or giant rattlesnake is popular. There is even a fiesta parade device resembling a Chinese dragon that local children walk inside of it guiding it along during the annual Children's Parade for the Taos Fiestas in July. Local artist Ted Egri has helped maintain the Viboron in repair for use by kids for the parade.

Then there is a little known legend about *el pichucuate*, a snake that seems to be associated with rivers and creeks. The *pichucuate* is reputed to be a small viper with horn-like shapes on its head above the eyes, very similar to some species of sidewinder rattlesnake. But the *pichucuate* does not have a rattle at the end of its tail. Not many contemporary New Mexicans have ever heard of the *pichucuate*.

One story was told by the late Max Padilla of Springer and Taos, New Mexico, about a warning he heard as child. He had said, "When we were kids and our elders knew we wanted to wander and play along creeks or *acequias*, my uncle Nestor would warn us. He'd tell us to be careful the *pichucuate* didn't get us. We understood it was a kind of snake that lived near water and that it's bite was supposed to be poisonous."

When asked what it was supposed to look like, he replied, "It was described as fat and grey slate rock colored."

Is this another of those folk tales used to warn children away from the possible danger of drowning, and or to be aware of venomous snakes? Is it possible in this Land of Enchantment a little known "horned" viper might have existed long enough to become the subject of speculation and folklore? It might be possible *pichucuate* is a regional Indo-Hispano name describing a sidewinder or similar snake with brow ridges. As far as fat and slate grey, there are water snakes that match that description. It still remains to be proven if it is only the stuff of legend or if there is in fact a rather scarce species of undiscovered snake that prefers riverine environments. Old timers who might know the answer continue to pass away, taking their knowledge with them as they leave this plane of existence.

Paul Gauguin's Mystery Bird

Michel Raynal

In Hiva-Oa, one of the islands of the Marquesas archipelago, in the middle of the Pacific Ocean, French painter Paul Gauguin and Belgish singer Jacques Brel spent the last years of their lives, and their graves can be found on that island. But cryptozoologists should also know it, as well as people fond of art, for in it dwells a bird unknown to science.

The Reports

On July 15, 1978, the first French TV channel TF1 broadcasted a documentary entitled "La Croisière de l'Eryx II" (The Cruise of the Eryx II). This charter-boat was said to have been loaned by a Brazilian man who wanted to go to Hiva-Oa, in French Polynesia, and catch a mysterious flightless bird said to live in that island. I do not know how he was aware of this rather obscure file (at least, in 1978)—he possibly knew some of the following accounts. One thing, at least, is sure: he did not find the bird.

Prof. Dante Martin Texeira, the curator of birds at the Museu Nacional in Rio de Janeiro was not even aware of the expedition. Michel Feuga, the author of a book on the cruises of the boat (1978), does not allude to this story. I would like however

115

to contact this Brazilian man, as elusive as the bird he was searching for, to discuss the latest news about this file...

In 1937, Norwegian explorer Thor Heyerdahl, who, 10 years later, became the famous hero of the Kon-Tiki expedition—through the Pacific on a balsa raft—was on honeymoon with his wife on Fatu-Hiva, Marquesas Islands. While visiting the nearby island of Hiva-Oa, Heyerdahl was riding in a mountain forest, together with a Polynesian native named Terai, when an incident occurred, which he relates in his book *Fatu-Hiva: Back to Nature* (1974):

> "Suddenly Terai halted his stallion and pointed to the trail in front of him. We had reached a hillock with low ferns, and a bird without wings was watching us. Then it ran faster than a little hen along the trail and disappeared like lightning into a sort of tunnel between the dense ferns. We had heard of this wingless bird, a strange species quite unknown to ornithologists. The islanders had often seen it but had never managed to capture one, as it always disappeared into holes and tunnels at great speed. Wingless birds in the Pacific are known only in New Zealand, represented by the kiwi and the now extinct twelve-foot-tall moa-bird. We searched a labyrinth of galleries criss-crossing the fern-clad hillock, but we failed to see this mysterious bird again."

An apterous bird in a Pacific island is not extraordinary: in the island fauna, many cases of apterism (partial or total atrophy of the wings) are known, among insects as well as among birds. In the Pacific ocean, one can mention the casoars of New Guinea, the kagou of New Caledonia (*Rhinochetus jubatus*), the Galapagos cormorant (*Nannopterum harrisi*), and the rails of Polynesia. The phenomenon is not limited to New Zealand, contrary to Heyerdahl's statement.

New Zealand, as noticed by Thor Heyerdahl, seems to be a paradise for apterous birds, with the kiwi (*Apteryx*), several rails, including the takahe (*Porphyrio mantelli*, formerly *Notornis mantelli*) and even a parrot, the kakapo (*Stringops*).

Recent bones, sometimes even not mineralized, have revealed the existence in New-Zealand of several flightless species of the Dinornithidae family (moas); some of them were hunted by the Maori natives some centuries ago, the largest one, *Dinornis giganteus*, looking probably like an emu 3.50 m high.

In 1980, I was able to contact Thor Heyerdahl, and 43 years after this sighting, he did not remember exactly the appearance of the bird, but he wrote to me nevertheless:

> "My memory of the wingless bird of Hiva-Oa has left me with the impression that it was considerably bigger than a sparrow and rather the size of a long-legged sea gull.
>
> "To my knowledge nothing else has been published about this bird although some ornithologist might have visited the island of recent years."

Nothing else published? Certainly not, as in 1956, French explorer Francis Mazière also heard of this bird on Hiva-Oa, as he writes in his book *Mystérieux Archipel du Tiki* (1957):

> "In Hiva-Oa Island, I heard from an old Norwegian sailor and from some natives of a very curious animal, named Koau, which looked like a kind of Kiwi.
>
> "According to the old Lee [sic] who had made after one on horseback but failed to capture it, as it was running too fast, the animal was as big as a cock, its fur [sic] was purplish-blue, its beak yellow, as well as its long and strong legs. It had only stumps of wings.
>
> "This description sounded so extravagant that I did not pay attention to it, until I came across a New Zealand magazine, relating, with photographs in support—which allowed us to show them

to the natives—the discovery, by a mountain expedition, of an unknown group of *noctunis* [sic]
which had taken refuge on the edge of a glacier.
The animal was the same. What should be emphasized is that, being unable to fly, this species could
have been transported only by migrations in canoes.
Unless, of course, the famous continent of Man
[sic] did really exist in the Pacific area."

Several errors should be noticed in Mazière's text, which
were certainly made by the publisher rather than by Mazière
himself:

1) The "old Lee" is Henry Lie, a Norwegian sailor living in
   Hiva-Oa, and a friend of Thor Heyerdahl's.
2) A "fur" is a curious integument for a bird. New Zealand
   kiwis have feathers looking like hairs, but in fact it is a
   misspelling for "plumage," both of these words being
   similar enough in French (pelage and plumage).
3) "Noctunis" is also a misspelling for "Notornis," as re-
   marked by Gabriel Lingé (1972).
4) Finally, the so-called Pacific Atlantis or continent of Mu
   (invented by Colonel James Churchward) should not be
   mistaken with the Isle of Man, between Great Britain and
   Ireland!

In 1977, I informed Bernard Heuvelmans, the father of
cryptozoology, of the previous reports. He forwarded them to
ornithologist Jean-Jacques Barloy, who mentioned them in his
book *Merveilles et Mystères du Monde Animal* (1979). Barloy
added some details on the mystery bird given to him by Francis
Mazière himself :

"The species is now extinct, the victim of a
foolish overhunting. Frenchmen knew the bird
and hunted it. Bones are said to be in some
tombs."

With regards to the name *koau*, reported by Mazière, it means "peak" in Marquisian, according to Dordillon's dictionary (1904), but the similar sounding name *koao* is translated as "a kind of bird."

More information on the koao is given by doctor Louis Rollin in his book Les Iles Marquises (1927):

> "KOAO: burrow bird which lives in the mud of the plantations of 'ta'o' [taro]. At the slightest noise, it burrows a hole, which makes its capture difficult."

This bird is also mentioned, curiously enough, in a linguistic study by Henri Lavondès, on the concept of hot and cold in Polynesia. In Marquisian language, the word *kena* means *burning*, and is used mainly to speak of the heat of an oven when the stones are red. In this case, another expression can be used, *matakoao*, or "eye of koao", and Lavondès writes:

> "The koao is described by informants as a bird with the size and looking of a duck, and characterized by its large red eyes."

Many rallids have red eyes, and the size of a duck is consistent with the bird "as big as a cock" described by Henry Lie.

Is it the Spotless Crake?

In 1979, Jean-Jacques Barloy suggested that this bird might be the spotless crake (*Porzana tabuensis*). This is a rail living in many islands of Polynesia, including Marquesas Islands. It is 15 to 20 cm (6-8 inches) long, black, with the rudiment of a tail, stumpy wings, red eyes and feet, and a light brown bill; it runs very fast, and would rather run from danger than fly. In the Marquesas, it would survive only in some valleys of Ua-Pou and Fatu-Hiva.

Jean-Claude Thibault, whose ornithological expedition to the Marquesas in 1973 failed to observe that species, wrote:

> "Several Marquesians assured us that the bird, when seen, burrows a hole in the mud and dives into the taraudières [tapped-out passages]."

It is exactly, and almost in the same words, what Doctor Rollin wrote about the koao in 1927.

Even the very native name of the bird supports Barloy's hypothesis: according to Holyoak and Thibault (1984), *koao* is the Marquesian name for the spotless crake. However Barloy (1979) had some reservations about his own hypothesis that this small rail might be the mystery bird sighted by Heyerdahl and Lie:

> "Is that species the koau? Its ecology and its behaviour are similar, but its size is much smaller. The mystery remains."

Let us remember that Heyerdahl's bird was "the size of a long-legged sea-gull," Lie's one "as big as a cock," and Lavondès's koao "the size of a duck," with some resemblance, leading to suppose that it possesses a strong beak, different from the thin, long beak of the spotless crake. Heyerdahl's mention of a "long-legged" bird confirms Henry Lie's statement that the bird he saw had "long and strong legs," quite unlike that of the spotless crake. In addition, the black color of the spotless crake much differs from that (purplish-blue) of the unknown bird.

If *koao* is really the name of the spotless crake, the confusion may have been introduced by Mazière: having recorded Lie's observation of a wingless bird, he may have asked the natives if they knew such a bird, and they spoke to him of the spotless crake, alias koao. Unless, of course, that the same word *koao* is given to both of these flightless birds by the natives: the concept of species among "primitive people" is actually quite different from that of taxonomists, as analysed by the father of structural anthropology Claude Lévi-Strauss, which seems to be confirmed by Lavondès's description of a koao differing from the spotless crake.

Is it the Takahe, the Moho,
or an Unknown Endemic Rail?

In an earlier paper about this file published in 1981 (my first cryptozoological article!), in the *Bulletin de la Société d'Etudes des Sciences Naturelles de Béziers*, I concluded that the mysterious bird of Hiva-Oa was very similar to the takahe from South Island, New Zealand, the discovery of which deserves to be recalled.

About 150 years ago, Maori natives of North Island of New Zealand reported the recent existence of a flightless bird, which they called *moho*, distinct from moas and kiwis: they had hunted it to the point of its extinction.

In 1847, Walter Mantell obtained some subfossil bones of moho from Waingongoro (Nord Island), which he forwarded to anatomist professor Richard Owen, in London, precising that the bird was also called *takahe* in South Island. It was a large rail, with rudimentary wings, which Owen described as *Notornis mantelli*.

In 1849, seal hunters were on Resolution Island, off South Island. Their dogs caught a big flightless bird: its plumage was purplish-blue, somewhat green on the back and the wings, it had a red thick bill and red strong legs. Thanks to Walter Mantell, the cadaver of this animal reached London, where it turned out to belong to the same species as the one examined by Owen. Supposed to be extinct from North Island before being discovered alive, the *Notornis* survived in South Island!

In 1850, a Maori caught a second takahe on Secretary Island, in the same area. In 1879 the dog of a rabbit hunter caught a new specimen, near Lake Te-Anau. It appeared that they were two different sub-species: the moho of North Island (*Notornis mantelli mantelli*), only known in a subfossil state, and the takahe of South Island (*Notornis mantelli hochstetteri*).

A fourth specimen was again caught by dogs in 1898, in the surroundings of Lake Te-Anau. The takahe was then considered extinct, although sightings were recorded in the first half of the twentieth century. In 1948, Geoffrey Orbell's expedition redis- covered this bird near Lake Te-Anau; it is now protected by New Zealand wildlife services.

In 1981, when I published my first article on this problem, the identikit-picture of the mystery bird from Hiva-Oa made me think of the takahe (*Notornis mantelli*) from South Island, New Zealand.

All the data were similar for both of these birds: ecology (habitat of forests and high grass of mountains), ethology (they ran to hide at the slightest alert), size (of a cock or a sea-gull) , rudimentary wings, strong legs, color of the plumage (purplish- blue), and same color of the bill and legs.

I also noticed the rapid run of the takahe, which allowed its incognito status until 1948: only four specimens were known previously, of which three ones caught by dogs, alone able to follow the bird in its habitat. This accounted well for the same elusiveness of the bird from Hiva-Oa.

Even the good taste of the flesh was shared by both of the birds: it was the cause of the extinction of the moho, New Zealand relative of the takahe, and of the rarity (if not the re- cent extermination) of the mystery bird.

Last but not least, Mazière said that his Marquisian infor- mants had recognized their mystery bird in photographs of "noctunis" (read *Notornis*)—irrespective of whether or not its native name is really koao.

There was just one little difference: the legs and beak of the unknown bird were said to be yellow, whereas they are red in the takahe. But these colors are very near in the visible spectrum, particularly if one remembers the conditions of observation, in the darkness of a forest, and the speed of the bird, which makes it difficult to examine at leisure.

I thus proposed, somewhat naively, that the Marquisian bird was the takahe (or the moho), brought from New Zealand by Maori natives in their dugout canoes. Incidentally, Gabriel Lingé made the same hypothesis in his book *Nouvelle-Zélande, Terre des Maoris* (1972), after he quoted Mazière and gave the correct spelling of the so-called "noctunis":

> "An unexpected argument which could support the hypothesis of early contacts between New Zealand and the islands of Eastern Polynesia is the following one. A bird believed to be extinct for a long time, the *notornis*, has been found again— and is living—in only two known places: Marquesas Islands and New Zealand."

In 1981, I was not so well aware as I am now of the avifauna of the Pacific, and of island biology, especially of the phenomenon of speciation (i.e., how new species appear). So my prefered explanation was an "importation" from New Zealand, because, as explained by Mazière, the bird could not reach Hiva-Oa by flying (it is flightless), nor by walking (as there never was any land of Mu!)—nor by swimming!

But I neglected the possibility that a rail once reached Hiva-Oa by flying, when it was still able to fly, and that it later evolved into a flightless form, as it is the case of most island rails. It is a well-known evolutionary process, which one explains by the absence of predators in these islands... until man arrived, with domestic predators apart from himself (dogs and cats). I only alluded to this possibility at the end of my 1981 article, when I concluded:

> "One can also imagine that the koau is another species of *Porzana*, or even a rail of a still unknown genus."

Jean-Jacques Barloy again alluded to the unknown bird from Hiva-Oa in his 1984 book *Les Survivants de L'ombre*, and he was less sure that it was the spotless crake:

"M. Raynal prefers to refer the koau to another
rail, the Notornis or takahe from New Zealand [...]
Why not?"

In 1986, Bernard Heuvelmans listed "my" bird in his famous
check-list of the 140 animal forms still unknown to science, relevant to cryptozoology, and he mentioned my hypothesis, but
he also expressed his own, less restrictive opinion:

"It has been suggested that it is closely related
to the New-Zealand takahe, and thus a species of
Notornis (Raynal 1980-81). What can be put forward more safely is that it looks indeed like a rail,
but larger than the local *Porzana tabuensis*,
thought to by locally extinct and surviving only on
other islands of the same archipelago."

Meanwhile, according to Olson's work (1973), the genus
Notornis had become a synonym with *Porphyrio*. It is true that
the purple swamphen (*Porphyrio porphyrio*) looks like the
takahe in its shape, size (only somewhat smaller), its color (purplish blue), its behaviour, etc., and all which has been said for
the takahe can be said for the purple swamp hen as well.

Moreover, it should be stressed that the same native name
moho, given by the Maoris of North Island to the extinct cousin
of the takahe (*Porphyrio mantelli hochstetteri*), is often given
to the spotless crake (*Porzana tabuensis*) in many islands of
the tropical Pacific: *moho* in Tuamotu, *meho* in Tahiti, *mo'o* in
Atiu, etc. This clearly demonstrates a link between both of these
flightless birds in Polynesian culture: it would not be surprising
if the same word *koao* is applied to the spotless crake (*Porzana
tabuensis*) in the Marquesas Islands where it is still living (in
Fatu-Hiva for instance), and to the unknown bird related to the
moho in Hiva-Oa. *Koao* would thus mean something like "flightless bird" in Marquisian.

Meanwhile, Ross Clark, of the Faculty of Arts (Auckland,
New-Zealand), also suggested the former existence of a form of

the purple swamphen in the Marquesas islands, from linguistics. In an article published in Sweden in the *Transactions of the Finnish Anthropological Society*, Clark proposed that a bird similar to the purple swamphen (*Porphyrio porphyrio*, the proto-Polynesian vernacular name of which is *kalae*), should have existed in the Marquesas islands or in the Society islands, from which came the first inhabitants of Hawai. The derived name of *'alae* refers in these archipelagoes to similar large rails (*Gallinula chloropus*, *Fulicula americana*), whereas the genera *Porphyrio*, *Gallinula*, and *Fulicula* are unknown in Eastern Polynesia.

Recent Discoveries on Hiva-Oa

In 1990, English zoologist Karl P. N. Shuker summarized the affair in an article on the birds still unknown to science for the *Avicultural Magazine*. He mentioned my first hypothesis, recalled the sensation created by the rediscovery of the takahe in 1948, and concluded:

> "Who knows, perhaps some future ornithological investigation on Hiva-Oa may engender a repetition of history!"

This had in fact almost happened two years earlier, with the discovery by David W. Steadman, of the New York State Museum, of subfossil bones about 1000 years old, of a new species of rail, named *Porphyrio paepae*, in archeological sites of two islands of the Marquisian archipelago: Tahuata and Hiva-Oa.

They have been discovered in *pae-pae* (hence the name given to the species), platforms used as a base for various dwellings, often kitchen middens, sometimes religious sepultures; this confirms Mazière's information reported by Barloy:

> "The species is now extinct, the victim of a foolish overhunting. Frenchmen knew the bird

and hunted it. Bones are said to be in some
tombs."

Though the external appearance of this rail cannot be determined from its bones, it is clear that *Porphyrio paepae*, which
obviously looks like the takahe (*P. mantelli*), and the mystery
bird, are one and the same species: in particuliar, it is likely to
have been purplish-blue in color, with yellow, strong bill and
feet. My cryptozoological analysis of this file in 1981, despite
its errors, was not so far from the truth.

The discovery of *Porphyrio paepae* represents an eastward
range extension of the genus *Porphyrio* of 3,200 kilometers:
the purple swamphen (*P. porphyrio*) only reaches the islands
of Western Polynesia, whereas a species extinct at the end of
the 19th century (*P. albus*) lived on Lord Howe Island.

Another representant of the same genus is known by some
bones about 3,000 years old found in New Caledonia by
palaeontologist Jean-Christophe Balouet, of the Muséum National d'Histoire Naturelle in Paris. It might have survived up
to the end of the nineteenth century, according to Balouet, who
collected the following account from a Kanak:

> "The chief of the tribe of Kele (Moméa) men-
> tions [...] a rail about fifty centimetres high, which
> looked much like a purple swamphen (*P.
> porphyrio caledonicus*) by its colours, but with a
> more massive beak, a white spot under its throat
> and a more greyish tail. His grand-father, who told
> him this story, was himself the chief of the tribe
> during the kanak insurrection of 1878, and he used
> to capture these not very shy birds with a snare to
> eat them."

Steadman states that the genus *Porphyrio* has never been
recorded, either living, or fossil, in Eastern Polynesia. In fact,
Mazière (1957), Lingé (1972), Raynal (1980-81), Clark (1982),
Barloy (1984) and Heuvelmans (1986) had already alluded to a

link between the mystery bird from Hiva-Oa and the takahe from New Zealand (*Porphyrio mantelli*), but I can understand he was not aware of these rather obscure references. He later filled this gap, paying homage to the early works by Ross Clark and Michel Raynal (Steadman 1997).

On the other hand, ornithologist Lionel W. Wiglesworth, as early as 1890, mentioned the presence of "*Porphyrio sp.*" in Raiatea (Iles Sous-le-Vent, Eastern Polynesia), from observations not confirmed since.

Steadman has suggested that fossils of the genus *Porphyrio* will be discovered elsewhere in Eastern Polynesia. Apart from Raiatea, I propose to search in Tahiti, where James Morrison, boatswain's mate aboard on the Bounty, mentioned an unknown large bird in 1789:

> "[...] The mountains produce birds of differ-
> ent kinds unknown to us, among which are a large
> bird nearly the size of a goose, which is good food;
> they are never observed near the sea nor in the
> low lands."

This incident is of course less known than the *mutinyrie*, played by Clark Gable, Marlon Brando, and Mel Gibson as Fletcher Christian, and Charles Laughton, Trevor Howard, and Anthony Hopkins as the terrible Captain Bligh). Derscheid (1939) thought it was a true goose, but the description is too vague to be so accurate. What can be said, however, is that no such large bird is known in Tahiti.

The confirmed existence, even indirectly, of the mystery bird from Hiva-Oa and the checking of its presumed zoological affinities provide new evidence of the efficiency of the cryptozoological research, following the method defined by Heuvelmans in 1988: it is what I emphasized in an article with Michel Dethier, summarizing the affair for the *Bulletin Mensuel de la Société Linnéenne de Lyon*, in 1990.

A Pictorial Representation

The file was at the same point ten years later, when on June 23, 2000, I received from Tahiti an e-mail by Philippe Raust, of the Société d'Ornithologie de Polynésie, who sent me as a joint file (.PDF) a copy of the journal *Te Manu* of June 1999. *Manu* means "bird" in Polynesian, and *Te Manu* is the bulletin of this scientific society. In this issue, ornithologist Jean-Yves Meyer, writing as JYM, published an article on the subfossil rail *Porphyrio paepae* of Hiva-Oa. Meyer quoted my 1981 article for the *Bulletin de la Société d'Etude des Sciences Naturelles de Béziers*, and he mentioned the further discovery of bones of such a bird by Steadman in 1988. But Meyer was not aware of my more recent writings on this file.

I thus informed Philippe Raust of this web page, with an updated version of my article, and he replied:

> "Yes, I read the article on Porphyrio paepae on your site. It is indeed very complete.
>
> "The writer (Jean-Yves MEYER) of the scientific news had not all the elements at hand, maybe.
>
> "But what seemed interesting to me, in addition to your analysis, was the remark on Gauguin's painting said to illustrate the beast.
>
> "Gauguin was in the Marquesas, at Hiva-Oa, at the beginning of the century. Could have he seen the bird or could have he heard of it?"

This was rather puzzling, as Meyer did not mention Gauguin in his article, even less a representation of the bird by the French artist. Amusingly, in several articles which I had devoted to this bird, for instance in *Cryptozoologia* of June 1994, I had mentioned that Hiva-Oa was known for Gauguin's grave. It was in my mind, in order to situate the island, which is probably unknown to most people, even in France. I was far from realising that this remark was not anecdotal, but quite relevant to the cryptozoological enigma!

I thus used a search engine on the Internet, and rapidly found several web pages relevant to my search, particularly the Internet site of the Musée d'Art Moderne et d'Art Contemporain (MAMAC) in Liege (Belgium), showing a painting by the famous French artist, "The Sorcerer of Hiva-Oa" or "The Marquisian with the Red Cape," showing our bird.

A historical background is necessary. Paul Gauguin was born in 1848 in Paris, and began to paint from 1875 in the impressionist wave. He made a first sojourn in Tahiti in 1893, and after a brief return in France, he came back to Tahiti, where he lived from 1895 to 1901. He then went to Hiva-Oa, where he stayed up to his death in 1903. Although Gauguin made many paintings in France, his Polynesian period is considered much more typical, because of the richness of its colours and exoticism (not to mention eroticism) of its subjects.

The painting kept in Liege, rarely shown in art books on Gauguin, was made in Hiva-Oa in 1902, some months before he died. The scene takes place near a forest, and the same place can be found in another painting, also painted in 1902, kept in the Cleveland Museum of Art (USA), *L'appel* (The Call): all the elements are almost identical (rocks, trees, water, and even the two women). The man with the red cape, giving the name to the masterpiece, is found also in *L'incantation* (1902), again with the same site. The name "Sorcerer of Hiva-Oa," given to the painting in 1949, is quite justified. According to Guillaume Le Bronnec, such a man, Haapuani, living in Hiva-Oa in the time of Gauguin, is the model for the artist, which is confirmed by Swedish ethnologist Bengt Danielsson, who published a well-documented book on Gauguin in Polynesia. (Incidentally, Bengt Danielsson was one of the members of the Kon-Tiki expedition in 1947, with Thor Heyerdahl, one of the witnesses of the mystery bird.)

The painting shows in its right corner, a dog and a bird, obviously the mystery bird I was studying since 1980. Its size (about 1/5 of the man, therefore about 35 cm high, more than one foot) is "the size of a cock," as claimed by Henry Lie, Francis

Mazière's informant, or "the size of a duck," as mentioned by Henri Lavondès. Henry Lie also reported a "purplish blue" colour, shown on the painting (with a green pattern on the back), and a "beak yellow, as well as its long and strong legs," and it can be found again in Gauguin's work. This beak is strong, unlike that of the spotless crake, and a red spot on the eye is consistent with Lavondès' remark on the red eye.

The resemblance of the bird in the painting with the purple swamphen (*Porphyrio porphyrio*), but even more with the takahe (*Porphyrio mantelli*), is striking: one understands that when Mazière had shown photographs of the takahe to the natives, they said that it was the same animal! And it is clear that *Porphyrio paepae*, described in 1988 by Steadman from subfossil bones, was painted alive by Paul Gauguin in 1902!

Now, is the bird still living? The most recent sightings are several decades old, as far as I know.

If *Porphyrio paepae* is already extinct, we shall have only some reports and subfossil bones, as it was also the case of another apterous bird, described and drawn by Sir Peter Mundy when he visited Ascencion Island, in the Atlantic Ocean, in 1656. This bird was never captured, nor even observed since, but Kinnear suggested in 1935 that it might be an unknown rail related to *Atlantisia rogersi* from Inaccessible Island. This hypothesis was fully confirmed (again an ill-known success of cryptozoology!) when subfossil bones of a new rail (named *Atlantisia elpenor* by Olson) were discovered in the 1970s in volcanic deposits of Ascencion Island.

The species may be still alive, though very rare, however. So, if field cryptozoologists go to Hiva-Oa, they should use drawings of birds of the genus *Porphyrio* (particularly *P. porphyrio* and *P. mantelli*) and the reconstruction of *P. paepae* which I proposed in 1995, or the one made by Morant and Bonet in 1998, and show them to the Marquisians: this might lead to collecting new recent reports, and accelerate the discovery of a living specimen of the cryptic bird, if by luck it is still surviving. This method was carried out successfully by Indian ornithologist Salim Ali for the Jerdon's courser (*Cursorius bitorquatus*), supposed to have become extinct for a century, and rediscovered in 1986 in India. Hopefully, may it be the same for the (now less) mysterious bird from Hiva-Oa!

A correspondent of mine fond of cryptozoology, Joël At, suggested to use dogs if by chance the bird is still living. He noticed that several specimens of the takahe (*Porphyrio mantelli*) had been caught by dogs in the 19th century with dogs. And in Spain, the purple swamphen (*Porphyrio porphyrio*) was surviving only in Andalusia, but in 1989 Jordi Sargatal i Vicens managed to catch several specimens with Labrador dogs to acclimate these birds in Catalonia. This excellent idea by Joël At lead me to wonder about the simultaneous presence of the bird and a dog in Gauguin's painting, and I realized that the dog is biting the back of the bird!

The hypothesis which I thus proposed in 2001 in an ornithological journal, *L'Oiseau Magazine*, together with ornithologist Jean-Jacques Barloy and Françoise Dumont (the curator of the Musée d'Art Moderne et d'Art Contemporain in Liège, Belgium), was that in 1902, Paul Gauguin saw the capture of the mystery bird of Hiva-Oa by a dog (the sole animal able to track the bird in the bush), thus giving an impressionist contribution to Polynesian cryptozoology.

Acknowledgements

Many thanks for their kind help to Joël At (Carcassonne, France), Jean-Jacques Barloy (Paris), Carl F. Carpenter (Englewood, Colorado), Françoise Dumont (Musée d'Art Moderne et d'Art Contemporain, Liège, Belgium), Bernard Heuvelmans (Centre de Cryptozoologie, Le Vésinet, France), Thor Heyerdahl (Kon-Tiki Museet, Oslo, Norvège), Angel Morant Forés (Valencia, Spain), Philippe Raust (Société d'Ornithologie de Polynésie, Papeete, Tahiti), Karl P. N. Shuker (West Bromwich, England) and David W. Steadman (New York State Museum, New York).

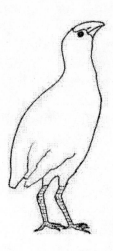

Author's reconstruction of *Porphyrio paepae* (1995).

Porphyrio mantelli (photograph supplied by author).

Spotless Crake, *Porzana tabuensis*
(photograph by Geoffrey Dabb).

Gauguin's painting (back cover) and detail of mystery bird
are reprinted with permission of the Musée d'Art Moderne et
d'Art, Contemporain (MAMAC), Liège, Belgium, (curator,
Françoise Dumont).

References:

Anonymous. 2000. Cryptozoologie: *Porphyrio paepae* (suite). *Te Manu*, n° 32: 6 (septembre).

Anonymous. 2004. Encore un oiseau mystérieux de Gauguin. *Ibid.*, n° 47 : 2 (juin).

At, Joël. 2000. Private communication (e-mail of 25 August).

Balouet, J. C. 1984. Les étranges fossiles de Nouvelle-Calédonie. *La Recherche*, 15 [n° 153]: 390-392.

Barloy, Jean-Jacques. 1979. *Merveilles et Mystères du Monde Animal*. Genève, Famot, 2: 115-117.

Barloy, Jean-Jacques. 1980. Private communications (letters of 24 June and 11 September).

Barloy, Jean-Jacques. 1984. Quels sont les animaux que l'on peut encore découvrir? *Trente Millions d'Amis*, n° 70: 27-32 (mai).

Barloy, Jean-Jacques. 1985. *Les Survivants de L'ombre*. Paris, Arthaud: 197-199.

Barloy, Jean-Jacques, and Pierre Civet. 1980. *Fabuleux Oiseaux*. Paris, Robert Laffont: 93-95.

Bhushan, Bharat. 1986. Rediscovery of the Jerdon's or double-banded courser *Cursorius bitorquatus* (Blyth). *Journal of the Bombay Natural History Society*, 83 [n° 1]: 1-14.

Bruner, Phillip L. 1972. *A Field Guide to Birds of French Polynesia*. Honolulu, Pacific Scientific Information Center: 43-44.

Clark, Ross. 1982. Proto-Polynesian birds. *Transactions of the Finnish Anthropological Society*, n° 11: 121-143.

Danielsson, Bengt. 1975. *Gauguin à Tahiti et aux îles Marquises*. Papeete, Éditions du Pacifique.

Derscheid, J. M. 1939. An unknown species—the Tahitian goose (?). *Ibis*: 756-760.

Dordillon, René Ildefonse. 1904. *Grammaire et Dictionnaire de la Langue des Marquises*. Paris, Institut d'Ethnologie de l'Université de Paris: 225.

Ehrardt, J. P. 1978. L'avifaune des Marquises. *Cahiers du Pacifique*, n° 21: 389-407.

Feuga, Michel. 1978. *Eryx II, ou la Croisière Polynésienne*. Paris, Robert Laffont.

Heuvelmans, Bernard. 1986. Annotated checklist of apparently unknown animals with which cryptozoology is concerned. *Cryptozoology*, 5: 1-26.

Heuvelmans, Bernard. 1988. The sources and method of cryptozoological research. *Cryptozoology*, 7: 1-21.

Heuvelmans, Bernard. 1996. Le bestiaire insolite de la cryptozoologie ou le catalogue de nos ignorances. *Criptozoologia*, n° 2: 3-17.

Heyerdahl, Thor. 1974. *Fatu-Hiva, Back to Nature*. London, George Allen and Unwin: 225.

Heyerdahl, Thor. 1980. Private communications (letters of 30 September and 17 December).

Heyerdahl, Thor. 2001. Private communication (letter of 19 April).

Holyoak, D. T., and J. C. Thibault. 1984. Contribution à l'étude des oiseaux de Polynésie orientale. *Mémoires du Muséum d'Histoire Naturelle, Série A, Zoologie*, 127: 1-196.

Kinnear, N. B. 1935. Zoological notes from the voyage of Peter Mundy, 1655-56. (a) Birds. *Proceedings of the Linnean Society of London*, 147: 32-33 (January 10).

Lavondes, Henri. 1971. Le chaud et le froid. Notes lexicologiques. In Jacqueline M. C. Thomas et Lucien Bernot, *Langues et Techniques - Nature et Société*, Paris, Klincksieck, 2: 395-403.

Le Bronnec, Guillaume. 1954. La vie de Gauguin aux îles Marquises (depuis son arrivée en 1901 jusqu'à sa mort en 1903). *Bulletin de la Société des Études Océaniennes*, 9 [n° 5]: 198-211 (mars).

Lingé, Gabriel. 1972. *Nouvelle-Zélande, Terre des Maoris*. Paris, Robert Laffont: 62.

Levi-Strauss, Claude. 1962. *La Pensée Sauvage*. Paris, Plon.

Maziére, Francis. 1957. *Mystérieux Archipel du Tiki*. Paris, Robert Laffont: 261.

Meyer, Jean-Yves. 1999a. Zooarchéologie et cryptozoologie: à la recherche des oiseaux disparus de Polynésie. *Te Manu*, n° 27: 7-8 (juin).

Meyer, Jean-Yves. 1999b. A la recherche de la poule sultane des Marquises (bis). *Ibid.*, n° 29: 4 (décembre).

Morant, Angel, and Carlos Bonet. 1998. El ave misteriosa de la Polinesia. *Biologica*, n° 25: 64-65 (octubre).

Olson, Starrs L. 1973a. A classification of the Rallidae. *Wilson Bulletin*, 85: 381-416.

Olson, Starrs L. 1973b. Evolution of the rails of the South Atlantic Islands (Aves: Rallidae). *Smithsonian Contributions to Zoology*, n° 152: 1-53.

Raynal, Michel. 1980-1981. Koau, l'oiseau insaisissable des Iles Marquises. *Bulletin de la Société d'Étude des Sciences Naturelles de Béziers*, n.s., 8 [n° 49]: 20-26.

Raynal, Michel. 1994. L'oiseau énigmatique d'Hiva-Oa. *Cryptozoologia*, n° 3: 1-8 (juin).

Raynal, Michel. 1995. The mysterious bird of Hiva-Oa. *INFO Journal*, n° 73: 17-21 (Summer).

Raynal, Michel. 2002. Une représentation picturale de l'oiseau mystérieux d'Hiva-Oa. *Cryptozoologia*, n° 47: 3-10 (janvier).

Raynal, Michel, Jean-Jacques Barloy and Françoise Dumont. 2001. L'oiseau mystérieux de Gauguin. *L'Oiseau Magazine*, n° 65: 38-39 (hiver).

Raynal, Michel, and Michel Dethier. 1990. Lézards géants des Maoris et oiseau énigmatique des Marquisiens : la vérité derrière la légende. *Bulletin Mensuel de la Société Linnéenne de Lyon*, 59 [n° 3]: 85-91 (mars).

Reid, Brian. 1974. Sightings and records of the takahe (*Notornis mantelli*) prior to its "official discovery" by Dr. G. B. Orbell in 1948. *Notornis*, 21 [n° 4]: 277-295.

Rollin, Louis. 1927. *Les Iles Marquises—Mœurs et coutumes des anciens Maoris des Iles Marquises*. Paris, Société d'Éditions Géographiques, Maritimes et Coloniales: 50-51.

Sargatal i Vicens, Jordi. 1992-1993. El retorn del gall mari. *APNAE* (Associacio d'Amics del Parc Natural dels Aiguamolls de L'Emporda), n° 1: 6-7 (hivern).

Shuker, Karl P. N. 1990. A selection of mystery birds. *Avicultural Magazine*, 96 [n° 1]: 30-40.

Shuker, Karl P. N. 1996. Mystery bird of the Marquesas. *Wild About Animals*, 8: 11 (September).

Shuker, Karl P. N. 1999. *Mysteries of Planet Earth*. Carlton Books Limited: 27-28.

Steadman, David W. 1988a. A new species of *Porphyrio* (Aves: Rallidae) from archaeological sites in the Marquesas Islands. *Proceedings of the Biological Society of Washington*, 101 [n° 1]: 162-170.

Steadman, David W. 1988b. Fossil birds and biogeography in Polynesia. In Henri Ouellet: *Congressus Internationalis Ornithologici* (19th: 1986: Ottawa), Ottawa, National Museum of Natural Sciences with University of Ottawa Press, 2: 1526-1534.

Steadman, David W. 1997. Extinction of Polynesian birds: reciprocal impacts of birds and people. In: Patrick V. Kirch and Terry L. Hunt, *Historical ecology in the Pacific Islands—Prehistoric Environmental and Landscape Change*, New Haven and London, Yale University Press : 51-79.

Thibault, Jean-Claude. 1973. Notes ornithologiques polynésiennes. II-Les Marquises. *Alauda*, 41 [n° 3]: 301-316.

Wiglesworth, Lionel W. 1890-1891. Aves Polynesiae. *Abhandlungen und Berichte des Königlichen Zoologischen und Anthropologische- und Etnographischen Museums zu Dresden*, 3 [n° 6]: 1-92.

"Dinosaur" Sightings in the United States

Nick Sucik

On a warm July night in 2001, three women of successive generations were driving along a county road near the rural community of Yellow Jacket, located 15 miles north of Cortez, Colorado. As they drove over part of the road where a creek undercut a culvert, an animal entered their headlights from the side. The driver braked, thinking it was a fawn deer. But once it was illuminated by the headlights, the two seated in the front stared briefly at an animal they had never seen before. Its body appeared smooth, devoid of fur or feathers. Its height perhaps was three feet and the small head was bent downward on a slender neck. The creature ran on two skinny legs with its tiny forelimbs held out in front of its body as it ran. Its body tapered down into a lengthy tail that, combined with the head and neck, made it about 5 feet long. The movement of the animal was noted as graceful, the head not bouncing as it ran.

The animal quickly passed in front of them and disappeared into the darkness. They were so awed by the spectacle that they both laughed at the absurdity of it all. One of them joked that maybe it had escaped from a local Jurassic Park.

Cortez, Colorado, is among the various areas within the United States that have produced sightings of creatures described by witnesses as remarkably dinosaur-like: appearing

137

both reptilian and bipedal. The subject has received little attention either in cryptozoological or Fortean circles for years.

Beginning in 2001, a small group of researchers pooled efforts
to begin investigating sightings of dinosaurs within the US, with
particular focus in Colorado. Our efforts centered primarily
around writing to newspapers and radio stations that served
regions where sightings had occurred, along with researching
folklore and indigenous beliefs of the target areas. During this
duration I had the fortune to conduct field research as part of a
cross-country journey I was taking after being discharged from
the military. This article includes the highlights of our team effort.
By providing this material I do not intend to make a case that
some relict species of dinosaur is persisting within the United
States. The animals described may not be dinosaurs or even reptiles for that matter, despite the labeling given by witnesses.
This investigation has only established that the phenomenon
of "dinosaur" sightings is not as isolated as previously thought.
In making this information available it is hoped that other researchers will be inspired to further explore the phenomena as well,
as there are a number of avenues yet to be considered.

Crosswicks, Ohio

In the late 1800s, a reptilian 'monster' was reported as having
attacked two boys, only then to be confronted by an angry mob.
The story was related by Hazel Brookes and published in a series
on regional folklore by Warren County Historical Society in Ohio.

As the story goes, two boys had been fishing on a creek near
the village of Crosswicks when a monstrous "snake" came
through the reeds and slithered directly at them. The boys ran
but the creature was able to catch one of them by "throwing"
out two long arms or forelimbs, while simultaneously producing two long legs or hind limbs, each of which were described
as four feet in length. (The description of the emergence of limbs
seems to imply that both sets somehow had been retracted or
otherwise held against the body before being deployed.) It then

began dragging the screaming boy toward a large sycamore downstream. The tree was described as 26 feet in diameter and hollowed out inside. Once there, the creature headed for a large opening on one side of the tree. The captive boy by then was so overcome with fear that he had ceased his struggles. Three adult men, who were working a quarry nearby and had heard the commotion, arrived on the scene just as the reptile was entering the tree with its prey. They shouted loudly and ran toward it, alarming the beast; it released the boy and hid inside the tree.

Later that same day, a party of five dozen men armed with axes and dogs surrounded the giant sycamore with full intentions to chop down the tree and dispatch its monstrous occupant. Amid the frenzied chopping, the creature suddenly emerged, threw out its limbs and raised itself up twelve feet and more in the air before dashing across the stream. The beast's ferocity and horrific appearance persuaded some of the men and dogs to forego pursuit, but the braver ones gave chase to the animal for nearly a mile before it escaped down a boulder-fortified hole in the side of a rocky hill.

The men described the creature being "from thirty to forty feet long, and sixteen inches in diameter, and the legs four feet long and covered with scales as was the body. Feet, about twelve inches long and shaped like a lizard's, of black and white color with large yellow spots. Head, about sixteen inches wide with a long forked tongue, and the mouth deep red inside. The hind legs appeared to be used to give an erect position, and its propelling power is in its tail." Some years later, a story emerged that three young men spotted a similar type of creature crossing a road near a swamp east of Lebanon, Ohio, but details were lacking.

As a personal bit of research, I spoke with a former resident of Warren County who had heard the tale from his grandfather. In the version passed on to him, a skirmish broke out between the men and the creature, resulting in a portion of its tail being chopped off before it could flee. The documented account above ends with the boy being hospitalized, but my informant said the boy ended up dying, after which the father had the tail portion destroyed.

Possibly related are several sightings of a large lizard-like animal in 1975 in Trimble County, Kentucky, not far from Warren County, Ohio. Investigator Mark Hall recorded witnesses' descriptions of a giant lizard with a red forked tongue, large eyes, and black and white stripes with quarter-dollar-sized orange speckles over them. One witness estimated the animal's length as around 15 feet. It also was described as a quadruped, but elsewhere it was stated that some witnesses had told Hall the animal ran bipedally. The coloration and noted spots match the descriptions given of the Crosswick creature, strongly suggesting a connection between the two, though the former had occurred 100 years prior and was not widely publicized.

In *Fossil Legends of the First Americans*, author Adrienne Mayors describes the origins of the annual Lizard Dance held by the Yuchi Indians. The Yuchi have been in Oklahoma since the early 1800s but originally had inhabited what is now modern-day Tennessee. The legend that serves as the basis of the Lizard Dance tells of three boys who were taken by a medicine man far from the village for special training in medicine knowledge. After they made camp in unfamiliar territory, the medicine man carefully scouted out the surrounding area. Upon returning, he told the boys to avoid a big hollow tree with a hole in its side. The boys later were sent for firewood and, as such stories go, one of them disregarded the medicine man's warning and attempted to chop wood from the hollow tree. A huge lizard emerged from within, seized the boy, then dragged him back through the hole into its den. The medicine man later killed the monster by leaving poisonous bait, after which they cut off its head and took it back to their village. It is from this event that the Yuchi celebrate with their annual Lizard Dance.

Mayors writes that, during the grand opening of the Oklahoma Natural History Museum in 2000, tribal members attending were excited at the display of a predatory allosaurid dinosaur, *Saurophaganax maximus*, which they exclaimed was the same as the monster lizard described in the legend. It is unclear, however, whether their reaction was because the dinosaur matched specific details passed down in oral traditions, or if it merely was a situation of defaulting a prehistoric monster for a legendary monster. Whatever the case, the similarities between the Crosswick story and the Yuchi legend cannot be overstated. Do both stories separately describe the same behaviors and habitats of similar beasts, or did the latter somehow inspire the former? Furthermore, whatever happened to the head of the monster which was said to have been brought back to the tribe? Interested researchers in or near Ohio may want to pay a visit to Crosswicks. According to my informant, his grandfather pointed to a spot where the monster had been encountered, suggesting that oral tradition of the event may still linger amongst the community.

Pueblo, Colorado

What set interest in motion concerning American dino-sightings started with an email sent to researcher Ron Schaffer. Ron was contacted by a visitor to his website who claimed to have encountered a small, dinosaur-like creature along the Fountain River near Pueblo, Colorado. This individual, 'Derick,' claimed that while dirt-bike riding with friends near the river they caught sight of an animal moving through a clearing. He described it as three to four feet in length, having a green body with black markings on the back and with a yellowish-orange underbelly. It walked upright on its hind legs, with its tail held above the ground. He also made out that it had two forelimbs or arms. He thought he saw a lump or horn above the ear. When the animal saw them, he claims it produced a high-pitched, birdlike screech.

He referred to the animal as a 'prairie devil' and implied that such animals were seen from time to time near the river. His email included a number of photographs which he claimed a friend had taken of the animals on separate occasions. However, the pictures were vague and far from convincing. The objects in the photos greatly hint of stationary models, especially given how close the photographer must have been to the 'devils' without alarming such small animals.

Chad Arment contacted Derick and informed him that, if the animals indeed were real, he would have to provide a better form of evidence. Several months later, Derick responded with a scanned black and white photograph of a man holding a rifle in one hand and what appears to be a dead dinosaur-looking creature in the other. The creature has a gunshot wound in its side. Derick's accompanying message stated that it had taken a while to acquire the image and that he was unsure when or where it was taken. The photograph is very impressive and, as Chad later would write in an article for *North American BioFortean Review* (NABR), there is no mistaking of the animal for any known species. From all aspects, either it displays a true unknown reptile and potential dinosaur, or it is a fake. Aside from the vague details surrounding the photo, Chad pointed out certain problems with the way the creature appears. Its posture seems too firm, with the tail bent inward and its mouth open, as opposed to the limpness that afflicts any animal recently dead.

Email contact with Derick was eventually lost and, as his last name was never given, there was no method of reaching him. This meant that if one took more interest in this mystery there was little else to do but personally visit Pueblo, Colorado, and use the few locations and references mentioned in Derick's emails to try to find the exact spot. In 2002, I took a trip into Pueblo and was able to locate the exact area Derick's sighting had taken place and where the photographs were taken along the Fountain River. After looking through tracks left along the river, I made random inquiries among residents living near the river. Many expressed interest in the matter but no one had any familiarity with anything called 'prairie devils' or even knew of any sightings involving large biped reptiles. One possible exception was a man who recalled that as a teenager he had heard of small dinosaurs (which he likened to the Composaurs featured in Jurassic Park) being seen on Bacalite Mesa on the outskirts of Pueblo. He was under the impression they were less than a foot in height which, if they were not fully bipedal, could suggest they were of a well-known species of semi-bipedal lizard such as the collared lizard. Access to Bacalite Mesa required special permission and, as time was a factor, I decided to bypass a look atop the Mesa.

The hunter photo Derick sent remains a mystery. That details of the photo's origins were lacking and given the images he previously had sent were likely fakes, a hoax seems most probable. Yet, for a prank, the reptile is remarkably detailed and likely would have taken considerable time and expense to construct.

Adding to the mystery, Chad noted that friends of his in the reptile trade at one point had done business with an individual who collected Colorado species and subsequently had offered to obtain for them what he called "river dinos." Having no familiarity with such reptiles and also lacking the funds, they turned down the offer but the description indeed was similar to that given by Derick.

Pagosa Springs, Colorado

In 1982 Myrtle Snow wrote to the *Rocky Mountain Empire Magazine*, the Sunday newspaper supplement of *The Denver Post*, about encounters she had had with 'dinosaurs' over the course of decades near Pagosa Springs, Colorado. Her letter came in response to an article in the paper on the debate about whether or not dinosaurs were cold-blooded. She cited four incidents betweeen1935 and 1978 that involved what she referred to as 'live dinosaurs.'

In her first account, Snow claimed that as a little girl she and a friend found a nest in an outhouse that contained five newborn 'lizards' about 14 inches in length. Her friend said they were "snakes with legs" and her friend's mother was summoned to identify the creatures. Unable to do so, she told the girls to take them by the tail and throw them into a field.

Snow's next account is most remarkable and hinges on her credibility. If the event occurred, it is most unlikely to find confirmation. Even if it is completely fictional, at least it makes for a story worth retelling.

As her story goes, sometime near the late 1930s, a rancher in the Pagosa Springs area was losing lambs to an unknown predator. A shepherd was armed and given the task of watching over the flock. In a canyon some five miles from the ranch, the shepherd allegedly shot and killed a strange creature. A young Apache ranch hand was assigned to haul the body back to the ranch using a train of mules after propping the carcass onto a makeshift Indian sled. The carcass was brought to a barn where area farmers came to view it the following morning. Myrtle, who was a child at the time, explained that her grandfather took her there and that she was able to see it up close. In her letter to the newspaper she described it as having been "about 7 feet tall, gray in color, had a head like a snake, short front legs with claws that resembled chicken feet, large stout back legs and a long tail." During my interview with her she added that its body was covered in fine gray hairs.

When I asked her what others thought the animal was, she said only the Apache boy seemed familiar with it. He told the rancher owner that it was what elders on the reservation called a 'Moon Cow' and they had once been seen from time to time but since had become rare and, if any of the children should see one, they quickly were to tell an adult. (In such a context, if that name sounds absurd, it should be kept in mind that for symbolic reasons the horse once was referred to as a 'Sky Dog' by the Apache). Myrtle thought the body was packed in ice and sent by train to the Denver Museum

Either that same year or a few years later, Myrtle recalls another sighting she had while visiting a friend's rural home. She had gone exploring by herself into the woods and she came upon what she called a den. As she examined the den, she heard a hissing sound causing her to stand still as she carefully looked about the ground for a snake. When she moved again, the hissing resumed. Then she realized that the sound came from a tree line where she saw a creature similar to the animal displayed in the barn. It was standing upright and reaching for something out of a tree with one of its forearms. Myrtle said it was green from the waist up but its bottom was half-covered in what looked like brown "hair." Running down its back from the neck were odd humps. Myrtle said that the animal plucked something from the tree and held the object for a moment as though examining it before 'plopping' it into its mouth. Someone called her from the house and she left the scene.

The next occasions that Myrtle claims to have seen bipedal dinosaurs occurred in 1978 and another a few years after her letter was published. In both of her accounts, she was in a vehicle being driven by someone when she saw a dinosaur. In the 1978 sighting the creature was seen walking along a fence during a rain shower. Both of these alleged incidents occurred within a few miles of each other on the same stretch of highway. In the later sighting, she alleges that the driver also saw the animal.

What is perplexing about Myrtle's stories is that she cites actual people who supposedly were involved in the incidents she describes. However, most already have died or have moved away without her knowing their present locations. Some died only a few years before I'd contacted her. I was able to speak with surviving kin in most cases but no one I contacted ever recalled hearing their deceased relatives mention anything about a dinosaur or giant lizard or monster of any sort. A cooperative archivist at the Denver museum took time to search through indexed listings of incoming items to the museum but he could not find any indication that a strange specimen was sent to them during the year Myrtle claims. That's not to say that, supposing the incident had occurred and the body was loaded onto a train, it wasn't sent somewhere else.

None of Myrtle's leads went anywhere, though bad luck or poor timing cannot be ruled out alongside more pessimistic assumptions. It did seem that Myrtle genuinely wanted to locate people who could back up her claims and she wrote to me offering suggestions as to how some individuals might be located. At any rate, it was not exactly an attitude one would expect from someone who knowingly fabricated a story.

Even if Myrtle's tales were utter fiction—contemporary dinosaurs lurking in the forested outskirts of rural communities— she inadvertently led to findings in an area where contemporary dinosaur sightings are current, frequent, and even have a hidden history.

Cortez, Colorado

Once while in the Marines, a friend and I passed time on mess duty by discussing cryptozoology. In the conversation, he mentioned something he had seen on a television show, either *Unsolved Mysteries* or the show *Sightings*. The program featured a segment involving a couple of ranchers from somewhere in the Southwest who during a horseback ride through the mountains came upon the body of what they took to be a small dinosaur. They preserved it and tried to notify regional academics, but no one bothered to look into the matter, with the exception of one individual who eventually bought the carcass from them. As such stories so often end, the person and the carcass 'never were heard from again.'

As my friend recalled, the segment featured a photo taken of the dead creature. At the time, I only had heard vague allusions to dinosaurs in the US and, given their rarity, I thought the story was worth remembering. After Chad compiled an overview of dinosaur sightings in the US for an article in NABR, my interest in the matters grew and I wrote to *Unsolved Mysteries*, in addition to posting inquiries on fan-sites for that series and also for the show *Sightings*. No further information ever came to light, nor could I find anyone in either of the fan forums who recalled such an episode. As a shot in the dark, I had asked Myrtle if she ever watched *Unsolved Mysteries*. Every day, she responded. Had she seen an episode featuring a small dead dinosaur, by chance? She said she had and that the story took place near Cortez, Colorado. I asked her what she thought of the segment, given how it seemingly would add credence to her own stories. Surprisingly, she did not think much of it at all and implied that the angle of the segment was not suggesting the dinosaur had been alive recently but more likely the ranchers merely had found a remarkably preserved mummified specimen.

Now, I still am no clearer that such an episode of either TV series exists, despite Myrtle's claims that she had seen it. But her pointing to Cortez as the source of the story did reveal a region that, for decades, has generated recurring sightings of dinosaur-like creatures, though it did not shed any more light on that mystery episode.

Hoping to learn something of this mysterious episode I wrote to *The Cortez Journal* and explained my interest in any information pertaining to stories or sightings of small bipedal, reptiles resemblant of dinosaurs. They ran the letter and received at least one response. A man contacted the *Journal* explaining that he had encountered a similar animal in California during the 1950s. His account, featured below, definitely was of great interest but unfortunately did not shed any light on the original mystery.

Despite that one intriguing report, the letter to the editor was otherwise unsuccessful. If the 'Dead Dino' episode had taken place in Cortez, then it seemed reasonable that someone would at least have known of the individuals it concerned. I was discouraged but decided to give Cortez one more shot. I placed a classified ad explaining the target of my interests, purposely vague enough so as not to inspire hoaxers, but revealing enough so that only someone with a relevant experience would likely know to what I was referring:

> *Research team looking for info regarding sightings of "two legged" reptiles, sometimes described as dinosaur-like. Confidential.*

The ad ran its course without a response during the weeks it was printed. At the time, I still was in Hawaii in the Marine Corps, expecting to be properly discharged within a short time. Only after I had given up on Cortez and, for the most part even had forgotten about the classified ad, did I receive a hit.

A young woman wrote a brief message saying that in the summer prior, near Yellow Jacket, she and her mother had glimpsed something that ran past the headlights of their truck which they thought might have been the two-legged reptile I referred to in the classified ad. It was from this brief message, which I didn't think much of at the time—expecting them to describe a sprinting collared lizard—that would come to set everything into motion.

A few weeks later I was honorably discharged from service and my car was shipped to California after which I was left to drive home to Minnesota. With half the country before me, I charted my path based on cryptid reports I wanted to investigate. When I arrived initially in Cortez, I only had that single account to go by. I met in person with the mother and daughter who had contacted me (their story being the account at the beginning of this article). When I interviewed them in the presence of a number of extended family members, it was immediately clear that the pair had never shared their experience with the rest of the family. I have no question about their sincerity. It was clear that they both saw something that night that they could not identify and that it left a deep impression on them. The mother even said that, after they arrived home, she and her daughter agreed to separately draw sketches of what they had seen, to confirm that they in fact had both seen the same thing. Their drawings matched and then both agreed to keep quiet about the matter. Neither had ever heard so much as a rumor of such a thing before.

Their account inspired me to invest a bit more time in Cortez. It seemed that I was going to be stuck with a one-hit-wonder as far as sightings went, unless I somehow came upon another account. The classified ad had only barely worked and one report would not be enough to get the local paper to bite. Even then, my two witnesses quite keenly wanted to avoid any publicity on the matter.

From there I was left without any other leads. The option left before me was to do a house-by-house survey in the area of the sighting but, after a few awkward attempts and one near-mauling incident with territorial dogs, I decided to change tactics. I reasoned that the most practical means of finding more leads was to check places that were likely to hear of such encounters: wildlife agencies, hunter clubs, and so forth. At Cortez's Reptile

Reserve, I had my second hit. The owner, Jeff Thurlin, was kind enough to listen to my strange pursuit as he worked. Afterward, I directly asked him if he ever had heard of anything similar from the region. He had.

Back in 1996, a woman who ran a local trailer park came to him claiming she had seen a bizarre reptile that looked "like a dinosaur." The woman, whom I will refer to as Bea, had seen the animal as it dashed past her front door. Bea thought it may have been coming from a pond close to the property and it is worth noting that this was during a drought year when water sources for wildlife were scarce. The sighting took place in April at around 11 am.

She described its dimensions as 3 ½ feet tall and just as long. The color was greenish-gray with no apparent markings. Its neck was maybe a foot in length with smooth skin. Its motion was described as graceful. As it ran, the body was streamlined with the tail sticking straight out, with the head and neck held forward. She said the legs reminded her of a frog's hind legs (which, during our interview, I understood as meaning they were thick closest to the body, becoming thinner in extremity). The tail she thought to be around 2 feet long and cone-shaped. Of the head, she caught little detail, except that it appeared to be fused with the neck with no distinguishable joint. She did not notice any arms or additional legs. It appeared to be supported entirely on two legs, which she referred to as its 'front' limbs.

Thurlin had searched through images of indigenous reptiles with Bea in an effort to find a match for what she had seen. No native species fitted and only a monitor lizard came somewhat close, though none are fully bipedal. Of those that can run on two legs, their posture is far from streamlined in the manner Bea described. Thurlin theorized that an exotic pet perhaps had escaped from one of the campers at the trailer park.

After recording her testimony, I told Bea that I had spoken with someone who, in just the prior year, had described seeing an animal of remarkable similarity. Her reaction was emotional, giving signs of immense relief. The sighting had haunted her. For almost a decade she was left without answers as to what she had seen or whether she really had seen anything at all. Bea's reaction to the event was common among other witnesses about whom I later would learn. The overwhelming majority found their experience to be isolating. It is one thing to see something akin to Bigfoot, whether one believes or disbelieves in the matter, as the concept is widely recognized. But seeing something that appears to be a dinosaur is rightly absurd by comparison. At the very least, some debate is possible over Bigfoot but, as everyone knows, dinosaurs have been extinct for 65 millions of years. This is not to say in fact that these sightings dealt with actual surviving species of dinosaurs. Other species resembling dinosaurs need to be examined, but the impression is so strong with witnesses that their choices are between coming forward and immediately being labeled crazy or to remain silent. Silence is the common reaction and we investigators are left with hidden phenomena that long have been overlooked and understudied, despite the fact that a fair number of sightings only recently have been uncovered.

Equipped with two sightings from Cortez, I notified *The Cortez Journal* of my findings in hopes that an article on the subject would summon further witnesses. The editor arranged for me to meet with a reporter who took down the information I collected and spoke to Bea over the phone. The initial mother and daughter witnesses agreed to let me describe their story but chose not to speak to the newspaper directly. After meeting with the paper, I left Cortez for further areas of cryptozoological interest in my journey cross-country.

The article was published a few weeks later, and not just in the *Journal* but also in its sister paper, *The Durango Herald*. Almost immediately on the day it came out, I began receiving e-mails from readers. The initial dozen or so mostly described sightings, usually from a car, with the overall majority clearly that of the same semi-bipedal collared lizards. This created something of a challenge. How was our target animal distinct from an exceptionally large collared lizard? With the two initial sightings, the size was far greater than the 16 inch maximum length that collared lizards are known to attain. Further, the postures described by Bea and in the Yellow Jacket account were different from that of a collared lizard and, lastly, the dino-type reportedly was seen holding its arms out to the front whereas a running collared lizard carries its forelimbs hanging at its sides. The dino-type also was described as having a long neck by lizard standards and its "graceful" motions seemed contrary to the shambling, side-to-side shuffle with which collareds run.

There also was the matter of size. It very much is human nature to mentally zoom in on an object during moments of excitement and thereby overestimate its actual size. Most descriptions of the dino-types' sizes were around 3 ½ feet and the reasoning I entertained was that if it was a collared lizard, it just as well could be ruled out by size. If any reports would qualify as the same creature from the two first cases, they would have to be clear cut enough that they were not of a recognized species or a misidentification thereof.

As indicated, some reports indeed described collared lizards. One from the '70s likely was an escaped emu (which were being raised in the area at the time). There were two separate reports that described in great detail a very strange little monster that stood on two legs, had sharp teeth, claws and a stripe of fur going down its back. In one case it produced a loud threatening hiss. Both included some very mammalian characteristics like fur and a coyote-like snout. On a subsequent trip to Cortez, I met with one witness who drew a sketch. Passing it along to Chad, he promptly identified it as a coatimundi, a member of the raccoon family. The behavior of the animal described by the witness and its physical description fit the coati like a glove. This answered another similarly described monster and a report that mentioned what appeared to be bipedal animal with large ears that was seen on a road. Such examples serve as an important reminder to researchers of how sightings of recognized fauna can evolve into near-fantastic descriptions.

There were three sightings that I received through readers that seemed to fall within our dino-lizard category. One took place in early July of 2002. In this instance the mystery prevailed by only a few inches as the driver had to slam on her brakes to avoid hitting the creature. It appeared to have emerged from cattails growing out of a drainage ditch along the road. Both she and her passenger watched as the animal, its head standing taller than the grill of the car, darted to the opposite side of the road. Though the event transpired too quickly to gather much detail, it at least gave an impression of size. For its head to be above the grill, it had to have been at least three feet in height, which would rule out a collared lizard. What could not be ruled out was a baby emu as such were being raised within the area. The witness explained that it had crossed her mind as a possible explanation, but then she recalled the animal appeared devoid of any beak or feathers or hair-like feathery features found on emus. Once they reached their destination they anncounced to the household that they had just seen a dinosaur cross in front of them.

Dove Creek, Colorado

It has been argued that dino-sightings are a recent phenomena inspired by the common presence of dinosaurs in popular culture nowadays. Yet sightings from Dove Creek, 35 miles north of Cortez near the Utah border, took place several decades prior, at the very least making a case that sightings in the Cortez area have a history to them.

A woman contacted me explaining how at her job there had been discussion of the newspaper article, to which one co-worker responded that he and his brother had seen the described animal during their youth. The man, who we shall refer to as Allen, had three separate stories, all taking place in roughly the same area, two of which occurred near a waterhole down in a small canyon.

As the story went, sometime in 1966 Allen, who was around 14 years old at the time, and his brother were making their way down to a waterhole. Abruptly, he looked up and saw a strange object maybe 15 feet away. It was well camouflaged within tall grass growing alongside the waterhole and, as it stood upright on its haunches, he did not mentally process that it was a live animal, momentarily wondering if it was a large toy of some sort. The animal appeared to have been eating something that it held in one tiny hand. After seeing the boys approach it stood perfectly motionless. Allen described the feeling that time had stood still as he and the creature stared at one another.

It was greenish-gray in color, which blended well against its surroundings. The body was devoid of feathers or hair. The mouth was round and he had the impression that it had teeth. He also noted the presence of a lump above the eyes. Its height was around three feet and it had an almost-human posture as it stood upon two well-built legs with rearward facing knees. What the legs had, the arms lacked. Seeming to extend out almost from the neck were two "dinky little arms." They were not attached to 'shoulders' as one might expect to see and their small size almost seemed to contradict the general form of the animal. In a way, Allen explained, the arms looked deformed, even hideous, yet cute at the same time, all adding a comical tinge to an already bizarre creature. When it finally turned and fled, it revealed a long tail with odd humps protruding from where the tail met the body. With the tail, he estimated its total length as being around 6-8 feet. As it ran away, it held its head and neck straight out.

Two additional incidents were cited. One occurred when the boys were playing in the waterhole and they heard something walking about in the rocks above them. Upon investigating, they only briefly saw a dark something sprint away, then found it had left behind three-toed, long-taloned spoors.

Allen's brother could not recall the first two accounts but he vividly did remember one incident that took place in the same area a few years later. He and Allen were looking for arrowheads when a strange creature about 12-14 inches high suddenly appeared from right behind him and scurried away at high speed. It looked like a miniature dinosaur but was moving away so fast he hardly was able to notice much fine detail as it kicked dust up behind it. However, as the animal ran away in a straight line, he did manage to see its back and noted what appeared to be one or possibly two rows of "spines" starting from beneath the head and running down onto the tail. From the few seconds he observed the head, it very much reminded him of an iguana. He could not recall the coloring exactly but thought it either was tan-brown or green-brown. The "front legs" were held in front with the fingers "curled" inward.

He was able to track the creature for a short distance. It had run into some brush but partway through there was an opening created by a red ant colony. Imprinted along the anthill were three-toed tracks, all about four inches apart from each other. The spoors were like those of a turkey but, whereas a turkey has toes that are spread out, these were closer together. He could not recall the imprint of a rear-facing toe.

For the most part, these descriptions did not seem to coincide with those described by Bea or from that of the Yellow Jacket creature. In elaborating on its leg structure, he said they appeared to come out from the sides of the body in a manner one might expect if an iguana was made to stand on its hind legs. In any case, the boys reported what they had seen to adults at their high school but apparently were ridiculed greatly. Still, they persisted that they were telling the truth and arrangements were made to have them speak with a teacher familiar with paleontology.

I will confess to caution about Allen's story of seeing the creature at close range. If taken at face value, it paints a fairly clear picture of a theropod dinosaur. Yet, he claimed to have heard of different sightings involving the same creatures over the years. In his own fashion, he had proven to be one of those 'too-good-to-be-true' type witnesses that most investigators encounter, in that the greater that person's claims, the higher the stake is raised for credibility. Still, his employer did vouch for Allen's sincerity. It is worth noting that some of the features he described complemented other reports. The strange eyebrow or lump over the eyes recalls the alleged Pueblo sighting that also mentioned a horn or lump above the eyes. Further, small humps along the tail sounds like Myrtle's mention of humps going down the back of the creature that she had seen reaching into a tree. Allen had mentioned the name of a man who lived near where their sightings had occurred and who had claimed, previous to the article's publication, seeing a strange type of reptile from time to time. Regrettably, during a follow-up visit to Cortez where through Allen I hoped to meet with the individual in question, I came to learn that Allen had tragically died the previous year.

Whatever it was that the two brothers saw, at the very least their stories evince that dino-sightings were not and are not recent phenomena. In point of fact, at one point in the 1960s the local museum even boasted that it had received the fresh remains of a 'baby dinosaur.'

'Baby Dinosaur' Mystery Skeletons
(Cortez, Colorado)

The overall response from *The Cortez Journal*'s readership was considerable and, when readers were not describing potential sightings, others would email theories as to what might be involved or to offer some guidance or even leads they thought could be helpful. Many said they intended to ask relatives or elderly residents if they knew or might have heard other stories. The amount of active interest was such that, for a time, it seemed that research on the matter was becoming democratized on a community level.

One individual wrote to explain that when he was growing up in the Mancos area near Cortez, there periodically were stories of sightings of a 'strange-looking, large bird-like creature.' It even was said that stories of these creatures could be traced back to a time when pioneers first settled the area. Sometime in the early 1960s, the carcass of one of these animals was brought to the local museum for display. It commonly was regarded that the animals had become extinct a couple of decades before. My contact suggested that I check through local newspaper archives around that time for more information.

Checking archives during a follow-up visit to Cortez proved a daunting task. Not only was the contact unsure what year exactly the carcass was displayed, during the approximate time period, Cortez had had not one but two separate newspapers. After two long days at the library scanning through microfilm, I finally came upon the article, which had proven to be an elusive hunt in itself.

In the June 13, 1963, issue of *The Montezuma Valley Journal*, the headline proclaimed "Cortez Museum Scores First with 'Baby Dinosaur' Skeletons" over a photo of the then-curator, Ed Roelf, standing beside a strange dark figure shaped like a small, elongated dinosaur on two legs minus any arms.

According to the article, two strange skeletons resembling "dinosaurs in the miniature" were found separately, one in a closed-off mineshaft and the other in a cave near the mine. The article detailed the curator's attempts to connect the bones to a species. Roelf had spoken to a Navajo man who said his people knew of the 'lizard-that-runs-on-long-legs,' but when asked what name they were given, the response was "Baby Dinosaur." Another Navajo from a separate area identified them as "hopping lizards," again implying tribal familiarity. Roelf eventually settled the question of identity by deciding that the skeletons were of a unique type of creature that had died out in recent years. The article hints at additional information Roelf presumably received from his informants, mentioning briefly that the creatures were thought to have been nocturnal and vegetarian.

I did some more research in hope of learning the fate of the bones. A woman from the Cortez historical society whose task it was to inventory unclaimed items from the community museum after it closed explained that a box had been found with the remains of the aforesaid "Baby Dinosaurs." The box was then shipped to the Denver Museum for examination and identification. Results there concluded that the skeletons actually were made from the articulated bone components of different species of mammals. In fewer words, Ed Roelf had been the victim of a hoax.

However this lead may have grounded out, it still failed to account for the previous history of sightings of strange birdlike creatures to which my informant had referred. I appreciated the likeness to birds, in that if we were in fact dealing with relict theropods, it would seem more likely their movements and/or appearance would be likened to that of a bird rather than a lizard. The Yellow Jacket witnesses likewise had noted the superficial bird resemblance, though the animal had forelimbs, a long tail, and was devoid of feathers. Given that the sightings dated back decades in time, it would seem plausible that the hoax was based upon a known and already ongoing phenomenon.

Big Boomers

The collared lizard or "Mountain Boomer" is given a maximum length of 16 inches. Seen running, the lizard is shorter than it is long and thus it is difficult to fathom how even a large male collared lizard could be mistaken for an animal 2 to 3 feet in height. The majority of sightings reported to me by people of the Cortez and Durango areas tended to describe something within such a height range. But certain other details almost assuredly would reveal a reptile to be a semi-bipedal lizard, presumably a collared. Those telltale details were posture, forelimb placement, neck length, and position of the rear limbs. Our target animal, based on three sightings that provided fairly clear-cut descriptions, had a posture more like a bird, arms that it held out toward the front or up against its body in front, a long neck and long limbs. In contrast, a Boomer's arms hang to its sides when it runs, it has a small neck, and appears to carry its body close to the ground as its rear legs move out from its sides, as opposed to erect and straight as with a bird.

If we segregate sightings that seemingly describe Boomers, we find ourselves with another mystery: how large can Boomers actually become? Certainly the reports received were describing something much larger than 16 inches in length, even if room for exaggeration is taken into account (as it always must). It is easy to shrug off an estimated size if, say, the circumstances involved a brief observation of something scurrying quickly across a road at night. It would be a separate matter entirely if and when a specimen actually is caught.

Not long after *The Cortez Journal* article on my dino-lizard research, a woman wrote to me about a 'Mountain Boomer' her three boys once had captured in New Mexico. The three were skilled in using their number in a game of tactical strategy to head off and eventually run down the speedy reptiles. On one occasion they came upon an exceptionally large, slower-moving lizard. Once they captured it, they saw that indeed it was a 'Boomer,' apparently an aged specimen, judging by its faint colors. The length of the body alone, not including the tail, was 20-24 inches (during our correspondence, she contacted her three sons, all adults now, and each independently confirmed the reptile had a body of at least 20 inches long). After thoroughly admiring the lizard, they released it.

A 'Boomer' with a body close to two feet long and sprinting on its hind legs certainly would catch a witness' eye as something phenomenal. As is sometimes the case, certain 'cryptids' can turn out to be sightings of unusual or unexpected behavior in otherwise known animals. While I'm confident from the sightings I've collected that something other than *Crotaphytus collaris* is responsible for the primary dino-lizard reports, hunting for evidence of a mega Mountain Boomer would be a worthy quest in itself.

Snowflake, Arizona

Quite by chance I came to learn of a sighting from Arizona involving a large dinosaur-like creature. Sometimes a sighting may surface in the most unexpected manner. While working on the Navajo Nation, a co-worker introduced me to a reporter friend who was visiting the area. My co-worker jokingly mentioned that I had spent time looking for living dinosaurs in the Southwest. Instead of finding the comment amusing, the woman reporter replied soberly that a close friend once confided that he had seen a "dinosaur" run across the road in broad daylight.

Eugene Atcitty was a staff member with the Navajo Nation's Emergency Management at the time of the sighting. The event occurred in the afternoon of February 17th, 1993, between three and four o'clock. He was the lead driver in a convoy of five vehicles heading for Phoenix for a Departmental Retreat. Within a mile of approaching a junction between Snowflake and Heber, Arizona, the highway took a sudden dip before climbing back to level ground. His vehicle had just climbed out of the dip, while the second vehicle entered it. It was after the climb out that he saw a large upright creature quickly crossed the highway into the large brush area. The height of the animal he placed at 10-12 feet. It had small arms that made him think of a T-Rex and its tail was carried above the ground as it ran. In moments the creature was gone from view. Traveling at approximately 55 to 60 mph, and being the lead vehicle, stopping was impossible. At a fuel stop in Payson, Arizona, a co-worker who was driving the third vehicle came up to him and quietly asked if he'd seen "that thing." It turns out that while the second vehicle was in the dip in the highway the third driver had caught sight of it as well. Both agreed to keep quiet about it for both fear of ridicule and concern that the creature, whatever it was, would end up being hunted and killed if learned of. So silence about this whole ordeal still stands, until now.

Some three hours travel from that location is Superstition Mountain. As the name would imply, it has its share of strange stories, one of which allegedly involved a dinosaur. I'm unfamiliar with the origin of the story or any details besides that a prospector and his mule were said to have been attacked by a "small T-Rex" type, bipedal dinosaur either on or near the mountain. As well, I also have heard from Bigfoot researchers about strange, three-toed tracks found on and near the mountain.

Big Bend, Texas

In an article for the now-defunct *Far Out* magazine, Jimmy Ward claimed to have heard stories of a "giant lizard that walked on its hind legs and whose voice sounded like the roll of distant thunder," said to live near the foothills of Big Bend National

Park. According to Ward, the creature was dubbed the "Mountain Boomer" by locals familiar with the stories, on account of the booming noise it made while running. The reptile was described as green or brownish in color, having powerful back legs and small, arm-like forelimbs, with hands comparable to a raccoon's paw. The rump and tail were said to resemble those of a kangaroo. All the stories Ward collected were admitted to be second-hand. He even expressed his own skepticism about its existence until he supposedly met a distressed family at a gas station and learned that they had seen a 5-6 foot tall upright reptile eating roadkill that then sprinted off with its tail held out straight as it ran.

As already shown, the term "mountain boomer" is usually applied to the collared lizard. That is not to say the term could not be used for more than one species as separate animals often are given the same name in different regions.

Ward died a few years after the article was written, leaving no means to confirm his stories. Yet there have been other allusions to dinosaurs in Texas. John Keel makes brief mention in one of his books of an incident where a "dinosaur" supposedly ran a car off the road somewhere in Texas in the early 1970s.

Mini-Rex

If dino-reports were to be clumped, they could form three categories based on size. Myrtle's monsters would fall into the first group where descriptions place the height of the animals as at least that of a man, if not more. The second group would pertain to smaller-sized specimens, described as around three feet in height which roughly summarizes the Colorado reports. The final category would be for reports of comparatively tiny creatures, around a foot in height or less. What keeps this last group from being dubbed lizards is their habitual bipedalness, in that sightings describe them as being on two legs at all times.

Before actually visiting Cortez, I sent a letter to the editor of *The Cortez Journal* explaining my interest in accounts or folklore surrounding strange, bipedal lizard-type creatures. Shortly after, the editor contacted me explaining that she had received a call from an excited reader who knew exactly to what I was referring and that the reader was anxious to share his experience. The editor put me in touch with the man to whom I will refer as "Jay."

Jay, a prospector in his mid-60s, explained that he had been all around the region, from Utah to Pagosa Springs, and had seen his share of wildlife. But he never had seen anything like I was describing in any of those places. Rather, what he saw was in northern California. He only told three people in his entire life about what he encountered around May of 1951. His family was traveling and they had pulled off the road to camp along the Russian River, north of the town of Ukiah. The area was forested heavily and thick with vegetation. Jay had taken a cooling dip in the river and was wading toward shore to retrieve his clothes when he came upon an animal so unique that, even though the event had transpired over 50 years before, he could remember it as vividly as if it occurred only yesterday.

The creature he saw was perhaps 7-8 inches tall but it stood completely vertical on its hind legs. The posture was more like that of a person than the way dinosaurs were depicted as being slanted forward. To Jay, it was just like a Tyrannosaurus Rex, but in miniature. The head he likened not to that of a lizard but to that of T. Rex, citing how ugly it was. Its head was not pointed but rounded and, in a manner very unlike that of a lizard, its teeth were visible while the mouth was closed. The tail was surprisingly stumpy and short. If the body was 8 inches tall, the tail had to be only 3 inches but Jay thought that, given its uprightness, the short tail probably was more than sufficient for balance. Its "arms" significantly were smaller than the legs. The entire body was a dull greenish-gray.

Both the animal and Jay stared at each other, neither moving until Jay decided to try to catch it. He had expected it to move in a clumsy manner but was surprised at the speed with which it fled into the brush. Jay knew he had seen something remarkable and apparently was mocked on the few occasions he had shared the experience with someone. In all his years of pros-pecting he said he had 'kept his ear to the ground' but never heard anyone else describe such a thing. Still, he was haunted by one nagging question: what if the animal he had seen was merely a baby?

Stories of mini-dinos will crop up now and then on cryptozoology forums and I have received a fair share of such reports from individuals who contacted me after having read about my research on the internet. Some of these can be easily explained as native species. Then there are reports in which someone not only will say they have seen a dinosaur but also are able to name the exact species of dinosaur they saw. Not surprisingly, velociraptors are a popular pick. While I tend not to give any weight to stories submitted online, or in forums and such, there is one exception that I have made so far.

One individual wrote to me after having read about my re-search saying it made him reconsider a story he had heard from his aunts. Their tale takes place between the Depression and the early 1940s when they would follow the crop harvests from state to state. His aunts, who could recall memories with clarity from those days, swore that during their travels out west they had found a "baby dinosaur" that to them greatly resembled "an itty bitty T-rex."

The critter would come to their camp when their mother was outside cooking and, during one such visit, the girls managed to catch it. It was kept in an old birdcage for a time and was fed any leftovers they may have had, taking both meat and veg-etables. It was described as having sharp little hooks on its hands and very sharp teeth, like that of a kitten. Its skin was like a lizard's but felt warm. It never tried to bite or scratch but it did not like being held. The animal behaved "like a tame squirrel." During the time they kept their pet, it grew from the size of a kitten to roughly the size of a cat, by which time it was far too big for the cage.

Because following the crops was a nomadic lifestyle, their father told them to leave their pet behind when it came time to move from the area.

The individual passing the story to me explained that he never exactly knew what to think of the tale despite both his aunts swearing it was true. He noted, however, that when the story first was told to him in the 1970s, they remembered that when it ran, it "flattened out, stretched its head out front, tail out back and was really fast." Such posturing was not widely accepted by paleontologists until years later.

The tale certainly has a romantic flavoring to it; tugging and teasing at probability once one gets past the irony: poor, uneducated migrant workers facing hardship merely trying to sustain themselves during the Depression, yet, in the midst of their toil, something comes along that is so incredible yet so natural that the magnitude of its significance is overlooked or simply goes unrecognized. In a way, as with Myrtle's story of farmers gathering around a dead theropod, such tales, whether they be true or not, add to the folklore of what our world would be like if dinosaurs lurked in secret and only stepped out into the clearing momentarily to be seen. If there were or are such things, they perhaps would possess the safety of being too unbelievable.

(Image by Aleksander Lovcanski.)

American Monitors

When conducting area-specific research into offbeat matters such as cryptozoology, and after one develops a reputation
for same, it is inevitable that people begin to seek you out as
someone with whom they can share their 'strange experiences'
with. Thus, while I came to the Four Corners area for the sole
purpose of researching the phenomena of 'dinosaur' sightings,
I became privy to a number of other strange stories. Some involved
other strange animals, some pertained to unusual or strange
occurrences such as livestock mutilations or UFOs, and there
even were some describing events seemingly of a supernatural
nature. One matter in particular did pique my interest, especially after it cropped up again and again in the different regions
I visited for dino-research. Over my course of time in the Southwest, I have spoken to three separate individuals who reported
seeing a type of giant lizard that failed the meet the descriptions of any native species.

The first incident took place only three miles south of Pagosa
Springs, in the same relative region where Myrtle Snow had her
alleged encounters. After writing to newspapers in the area in
an effort to locate other witnesses, Chad was contacted by
former police chief Leonard Gallegos. His sighting took place
sometime in the early 1950s when he was 10-12 years old. He,
along with his mother, his sister, and a brother, were on a fishing
trip in a fairly remote undeveloped portion of the San Juan
River. The four were walking toward a large rocky outcropping
with Leonard in the lead, when he suddenly spotted a large reptile perched on a boulder only 30-40 feet away. The animal was
at least the size of an alligator, he explained, though it was more
similar in appearance to a Komodo Dragon. Its color was greenish-
gray, with a snout similar to that of a horse and with a lizard's
body of at least 5 feet long. The reptile looked at them and began
bobbing its head. Leonard noted that it did not appear menacing
but the mother told the kids to flee. Later that day, his father
scouted out the area armed with a rifle but failed to locate the
animal.

Many years later he would hear of a Spanish woman who had seen a "dragon" near the area but did not get to speak with her before she passed away.

A giant lizard sighting was not exactly on the agenda for fishing around in Pagosa Springs for dinosaur reports but the account certainly was welcomed, if a tad confusing as to whether it related to Myrtle's sightings or not. Likewise, among the numerous emails I received following *The Cortez Journal* article was a message from a woman named Janet who wrote that she had seen a large, quadruped lizard-type animal near her home in Lebanon, 11 miles north of Cortez. The incident happened in the fall of 2001. She was driving toward home along a secluded dirt road when she spotted the creature apparently basking on the road ahead. It was about five feet in length and appeared dark brown in color. At the vehicle's approach, it lifted its body a few inches off the ground and scurried off the road in the direction of a creek. She commented that the way it ran straightforward reminded her more of a lizard's motion that did not resemble the side-to-side gait of an alligator.

While I lived on the Navajo Nation, I heard a number of allusions to giant reptiles, sometimes said to have horse-like heads. The matter is difficult to research, however, given that reptiles, and in particular snakes, are deemed taboo topics by tribal traditions. Furthermore, the appearance of anything strange or otherwise out of the ordinary often is regarded as a bad omen. For example, in the summer of 2002 there was excitement in the valley where I was living as someone had seen a lizard or a snake with the girth of a man's thigh. Later that summer, two boys drowned during monsoon floods. Tribal elders connected the two events, explaining that the appearance of the monster should have warranted a ceremony to avoid the misfortune it was foretelling. Hence, doing cryptozoology research out there is a bit more complicated than one would expect. Furthermore, the Navajo term used for 'giant lizard' is the same as that used for 'giant snake.' The only actual witness I spoke with on such a matter was an Anglo man married to a Navajo woman. Sometime in the mid-90s, he was walking along a dirt road when he spotted what seemed to be a huge lizard of a length that took up half the road. Later, he mentioned the incident to one of his in-laws who, with much excitement, replied that she too had seen a monster lizard but had been afraid to mention it to anyone.

That marks the only first-hand account I personally have collected about giant reptiles on the reservation. But it is worth noting additionally that a reporter from the main tribal newspaper mentioned to me that there is a particular spring where it is said that people saw a type of gigantic lizard that ran upright on its hind legs and was close to a man in size.

An additional indigenous reference can be found in Adrienne Mayor's *Fossil Legends of the First Americans* where she includes a traditional story amongst one of the Plains Tribes that describes a group of braves encountering a small number of giant lizards. No detail of the animals is provided but in the tale one of the braves decides to prove his courage by attacking one of the creatures and is killed in the process.

Exotic escaped animals obviously are candidates for such accounts but when sightings recur over extended periods of time (as they do on parts of the Navajo Nation), isolated cases of released or lost pets fall far short of answering all sightings. Morphologically, I remain curious as to why giant lizards seem to have horse-shaped heads in references made by Navajos, which incidentally echo the creature that Leonard Gallegos saw.

References and Rumors

There are still potential leads that have yet to be properly explored. As stated earlier, John Keel has mentioned in a few of his books about having received information on dinosaur sightings. Yet in none of his published works does he offer much more than a brief mention of these individual cases. Perhaps the matter seemed too abstract even for an experienced Fortean investigator as Keel.

Such was the case for investigator Christopher O'Brien, best known for his work on the mysteries surrounding the San Luis

Valley in southern Colorado. O'Brien caught wind of the Cortez article and contacted me. He explained how in either 1994 or 1995 after giving a lecture a state college in Alamosa, Colorado, a student told him his father had seen a dinosaur-like creature along the base of the Sangre De Cristo's Blanca Massif. The student said his dad described the animal as being upright, bipedal, about 3 ½ feet tall, with smaller front legs or arms. The man spotted the creature running along a fence while he was driving. O'Brien commented that at the time he had not followed up on the story due to its seemingly fantastic nature. It makes one wonder what other researchers may have picked up on a dino-sighting at some point but likewise shrugged the matter off?

In a letter to Chad, Gary Mangiacopra said that he recalled Ivan Sanderson mentioning on one of his talk programs that he was receiving reports from people from a Southern state, probably Alabama or Mississippi, of "turkey-size" dinosaur-like creatures. One of the members of our research group, Daniel MacCallister, mentioned there had been reports out of a part of Alabama during the mid 1990s where a 3-foot tall lizard or dinosaur-like animal had been seen chasing prey across a road. The matter was the subject for a local radio program at one point.

Sightings continue to this day. Quite recently I was contacted by a rancher near Dove Creek, Colorado who described seeing something at night that was quite unlike a collared lizard. Details are currently being gathered. If nothing else, this phenomena demonstrates how single sightings of strange creatures, or one-hit wonders, needed to be properly explored just as with any of the more familiar forms of cryptids—no matter how absolutely absurd the matter sounds.

References:

Arment, C. 2000. Dinos in the U.S.A. *North American BioFortean Review*, Volume II, Number 2.

Brookes, H. @1978. *Crosswick Monster*. Folklore Series, No. 11. Lebanon, OH: The Warren County Historical Society.

Clark, J. 1993. *Unexplained!* Detroit, MI: Visible Ink Press.

Coleman, L. 1990. Other lizard people revisited. *Strange Magazine* (5): 34, 36.

Cromie, D. 1963. Cortez museum scores first with 'baby dinosaur' skeletons. *The Montezuma Valley Journal* (Cortez, CO) (June 13): A1, 5, 6.

Hall, M. A. 1991. *Natural Mysteries*. 2nd revised edition. Bloomington, MN: Mark A. Hall Publications and Research.

Keel, J. A. 1994. *The Complete Guide to Mysterious Beings*. New York: Doubleday.

Ward, J. 1993. The mountain boomer. *Far Out* 1(4): 45-46.

Mayor, A. 2005. *Fossil Legends of the First Americans*. Princeton, New Jersey: Princeton University Press.

Back cover image shows an adult
Collared Lizard, *Crotaphytus collaris*.

Miscellaneous Menagerie

Chad Arment

What follows is a wide assortment of old newspaper articles noting reports of strange creatures around the world. Some are better known than others. A few are obvious out-of-place animals or outright hoaxes. The point of this exercise is not, however, to create any sort of circumstantial case for mystery animals, but to provide some notes on such stories and point to cryptozoological reports which would be worth further investigation.

Four Boys Catch Baby Octopus in Lake Erie

Mentor, O., (UP)—Four boys who went fishing Tuesday came home with the strangest catch of this or any other season in Lake Erie.

John Peakovis, 17, snagged a baby octopus, 23 inches long, and killed it with his knife.

Mentor harbor yacht club officials said they thought the sea creature had been brought north by one of the yachtsmen who winters in Florida.

—Edwardsville (IL) *Intelligencer*, May 24, 1950.

This, of course, follows nicely on the heels of the chapter on suspected freshwater octopuses found elsewhere in this anthology. I have heard stories of other saltwater species being found in Lake Erie, usually best explained as introductions via ships or bait buckets. One point to note, at 23 inches, the octopus mentioned was hardly a "baby." It could very well have been a mature adult, depending on species. Given that fishing on Lake Erie is a popular recreation, I would not give the possibility of lake octopuses much credence or attention, particularly without further corroboration. The Great Lakes, however, do have some intriguing cryptozoological reports that have barely been investigated—an extensive survey of mystery animals in these bodies of water would be very useful to cryptozoology.

Huge Octopus Seen Near Hawaiian Isle

Honolulu, Dec. 15 (AP).—Reports that a giant octopus with tentacles 20 feet long and a head three times the size of a man's, had been seen dragging itself along the rocks close to shore caused fishermen on Maui Island to cast their nets with caution.

Henry Sylva, son of a county official, reported sighting the monster and his story was vouched for by fishermen and swimmers.

Fishermen said the big devilfish had probably been driven in by high winds. Devilfish or squid, is used for food by Hawaiians and numerous big fellows were caught and taken to market during the rough weather.

—Syracuse (NY) *Herald*, December 15, 1932.

Now this is intriguing, primarily because Nick Sucik has already located reports of giant octopuses in Hawaii waters (*North American BioFortean Review* #5, available at StrangeArk.com).

This news account dates prior to the story Nick located in the July 27, 1936 edition of the Honolulu *Advertiser*. Now, this particular size estimate is well within the size range of the giant Pacific octopus, *Enteroctopus dofleini*, but that species is only known from north Pacific waters. Another octopus, *Haliphron* (provisionally identified as *atlanticus*), was found in New Zealand waters in early 2002. This deep-sea specimen was estimated at a length of 4 meters and weighing 75 kg when alive. Given that the Hawaii octopuses are only sporadically reported, perhaps they are another deep-sea species that rarely drifts into Hawaii's waters?

French Hunt Prehistoric Lake Monster
Man Sticks to Story of Seeing Scaly Beast With Phosphorescent Eye

Saint Gaudens, France, Sept. 3 (AP)—Hundreds of Frenchmen lost sleep and got wet in a vain hunt over the week-end for "the scaly monster of Lake Camon."

A violent storm swept over the Tartarins district as hunters, ready for anything, splashed about the reed-grown borders of the lake looking for a creature which had been described to them as a "prehistoric beast with phosphorescent eyes."

Nothing has come of the hunt up to now, but the determination of the hunters, amateur scientists and the curious grew as M. F. Narud, the man who reported seeing the beast, stuck to his original story and enlarged on aspects of the beast.

It was thick, and about 25 feet long, he said. It moved cumbersomely and an eerie light shown from its eyes, he stated. Its body, he believed, was covered with scales.

In the popular mind credence was given to the possibility that a prehistoric beast had survived these times when geologists came forward with theories that Lake Camon is a remnant of an ancient sea.

The strange hunt brought into action an official of an ancient office—the wolf lieutenant. In olden times it was his duty to protect districts from predatory animals. The popular opinion was that a prehistoric monster would certainly be classified as predatory.

Taking that view, Wolf Lieutenant Moga organized a search. He headed an "official" party which was aided by many individuals who came out in the rain to beat thru the reeds.

The hunting season has just opened, and many citizens are forsaking ordinary game for the hunt about the old lake.

—Dothan (AL) *Eagle*, September 3, 1934.

This article is more than a little suspicious. At this point, there appear to be some geographic anomalies in the article. I've not found, for example, any confirmation of a "Lake Camon." The description could very well point to an out-of-place crocodilian, as the French have their share of natural history enthusiasts and explorers, but unless other details emerge, I'm inclined to treat this as a newspaper hoax.

A Mokelumne River Monster

Upon the highest Celestial authority there is at the present time habiting the Mokelumne river, only a few miles from this place, a short distance above the Big Bar bridge between Jackson and Mokelumne Hill, a water monster that has spread consternation and terror among our entire Chinese population. They assert that one day this

week a number of them were on the river just below the old Boston mill site, when they saw what
they all supposed to be a large fish near the top of
the water, and one of their party made a leap into
the water to capture the prize. But no sooner had
he struck the water than the monster turned upon
its would-be captor, and instantly drew him under, from whence he had not arisen at last accounts. The monster is said to have teeth like a
dog, and to have powerful strength, as the man
was as nothing in his grasp. The Chinamen who
witnessed the tragical fate of their unfortunate
countryman, concluded that the thing was the
devil, and fled for their lives. It is also claimed that
another Chinaman lost his life in about the same
manner, near this place, several years ago, and it
is thought that this is the same monster returned,
if indeed he has not inhabited the Mokelumne ever
since. While we are not prepared to vouch for the
truth of this story, yet it is evident that something
has occurred or been seen on the river to excite
our Chinese denizens and put them ill at ease.
Some of our daring and curious citizens talk of
getting up a party to investigate the matter, and,
if possible, capture the monster, which they believe to be the Calaveras *Chronicle's* long lost
snake, which has not been heard of since Higby
went into the Internal Revenue Department.—
Amador Dispatch.

—Reno (NV) *Evening Gazette*, September 10, 1881.

This is a generic water monster story, not particularly useful in and of itself, but points to a specific place where some
historical research might be fruitful.

Alligator in Nevada River
Trapped After It Had Killed a
Number of Calves on Frey's Ranch

Winnemucca, Nev.—James Bryant and William Brennan, veteran Nevada trappers and hunters, arrived here recently with a tale of having trapped an alligator on the Humboldt river, near Frank Frey's ranch, west of Winnemucca.

Bryant and Brennan trapped the alligator after it had killed a number of calves on the Frey ranch and put the entire district in fear of going near the river. The trappers say it was an alligator that escaped from a circus train at Imlay a few years ago and has been living in the Humboldt river ever since. The alligator skin will be made up into a traveling bag by Frey. In addition to the alligator, the trappers secured a large number of muskrats, beaver, otter, coyotes and wildcats.

—Nashua (Iowa) *Reporter*, April 8, 1920.

No reason at present to suspect this is anything other than an out-of-place crocodilian, though the "circus train" story is certainly folkloric—which doesn't mean it can't be true, of course.

Winged 'Cat' on Ninth Life Is New 'Discovery'
Mystery animal in Washington Reported
to U. of C. Zoology Department.

Berkeley, July 9.—At last science is on the possible verge of a great discovery regarding the number of lives allotted to cats. An animal resembling a cat, except that it has wings with a spread of about one foot, has been discovered by Arthur Kingray, Washington State rancher.

Kingray, who reported his find to Dr. Joseph Grinnell of the University of California museum of vertebrate zoology today, thinks that the animal was a cat living its ninth life, and was so near to cat heaven that its wings were already sprouting. The creature had a cat's head and body, with four rows of muscled flesh having a spread of one foot down the back, and weighing about 25 pounds, the rancher informed Dr. Grinnell.

"If this man is not suffering from a hallucination, the creature he killed may be a flying fox," says the zoologist. "The flying fox comes nearest to fitting this description. It is a member of the bat family, comes from Southeastern Asia, and is noted for causing great destruction to fruit orchards."

Great pains have been taken to exclude this pest from the United States, and if it is slipping in, trouble lies ahead for the orchardists, Dr. Grinnell avers.

—Oakland (CA) *Tribune*, July 9, 1926.

Something is screwy about the figures here. A flying fox doesn't weigh more than 3 or 4 pounds. The description of the animal is extremely vague; perhaps investigation of Washington state newspapers would provide further details or a photograph. It may just have been one of those felines with a medical condition resulting in wing-like projections.

A Big Snake Story
The Wonderfully Strange Reptile
Discovered by Veracious Californians.

[Concord (Cal.) Sun.]

On Wednesday, learning through Thomas O'Neal of this town, that a party of wood-choppers, while felling a tree the day previous near the old Smith Ashley rancho, Pacheco, and some three miles from Concord, had discovered an enormous snake, the dimensions of which arouse our attention, we immediately repaired to the scene of the discovery. There, before us, lay an object, the body of which was snake-like in proportions, measuring fourteen feet seven inches long, three feet around the largest part of the body, and tapering gradually toward the tail. The head resembled that of a crocodile, a row of large, pointed teeth lining its jaws, the upper jaw extending forward from the anterior angle of each eye; the skin was light colored, spotted with black; a coat of hair, in places twelve inches long, extended in a continuous line from head to tail; and the body being ornamented with eight pairs of short webbed feet. The entire monstrosity would weigh about two hundred pounds. The discovery was made after the tree fell, which was found to be hollow, the reptile crawling from it, which so scared the wood-choppers that they fled to town, leaving hats and coats behind. Its general appearance was that of the amphisbaena, a genus of saurians that abound in South America, and which live mostly on small insects. But how this particular creature came in this vicinity is for the savants of California to study up. We are sorry the reptile was killed, as it might have adorned the interior of a menagerie. However, we should suggest that the officers of the California Academy of Sciences dispatch one of its experts to determine the species of reptile.

—Warren (PA) *Ledger*, December 25, 1885.

The reporter apparently created this explanation from thin air, as the animal described is nothing like an amphisbaenid, or any other known creature. I won't rule out the possibility that the true description was mangled by excited witnesses, but a hoax is the most feasible explanation.

Fish With Nine Legs.
Curious Thing Caught in Nevada
Goes to Washington Museum.

The New York *American.*

State Comptroller Sam Davis, Warden John Considine, Attorney-General James Sweeney and Louis Bovier, all of Carson, Nev., recently landed one of the queerest freaks in the fish line ever seen in Nevada. It is neither fish nor animal, yet closely resembles both. It measures about eighteen inches in length and has nine legs. It has a head, fin, gills and scales of the fish, but the nine legs are covered with fine, downy hair.

Mr. Davis and his companions were fishing with the regular flies when the animal-fish sprang from the water and hooked on Mr. Davis' line. For ten minutes a battle royal took place, the fish being finally landed on the bank nearly exhausted. As he was about to pick up the strange creature, however, it began to walk away. It was not until then that the party noticed that the fish had legs. So taken back were they that the strange creature came near escaping in a nearby alfalfa field.

Unfortunately, both for science as well as for the curiosity of the party, the animal-fish was killed by the excited men in their efforts to capture it alive. From one who claims to have seen the freak it is learned that it is to be embalmed and sent to the Smithsonian Institution.

—Syracuse (NY) *Post Standard*, August 31, 1905.

Again, an animal more fabulous than realistic. However, in this case, the witnessess truly existed. Sweeney was Attorney General for the state of Nevada during the 1906 San Francisco, California, earthquake. Unfortunately, no mention is made of where the creature was caught. In this case, I'm leaning more toward misidentification than hoax, though I cannot supply a reasonable biological candidate. Perhaps some intrepid investigator will have opportunity to make an appointment with the Smithsonian and try to track the critter down.

A Monster Gorilla.

Interesting information regarding huge gorillas of hitherto unknown species has been obtained by Eugene Brusseaux, a French official and explorer from northern Africa. One of these strange monsters was shot by one of the official's sharpshooters. The animal measured seven feet six inches in height, was four feet in width across the shoulders and weighed 720 pounds. One of the hands, when dismembered, weighed six pounds. It required the united efforts of eight native soldiers to drag the dead body of the beast from the point where it was killed to the French residency at Quessou, the administrative center of central Sangha. The animal was here skinned and buried.

Reports have been received at this station frequently during the last few months of the presence of these big creatures in the upper valleys of Loani and Sangereh, but hitherto it has been impossible to come to close quarters with them. According to native reports, however, the animals are unusually ferocious, not hesitating to attack caravans during their passage through the country.

These gorillas differ essentially from others. The ears are small, the shoulders and thighs are covered with dense and long black hair, while the chest and stomach are almost bare. It is believed that they belong to a species that has not heretofore been seen by white men.

—Lincoln (NE) *Evening News*, October 23, 1905.

The area in question is now the Central African Republic. Given that the mountain gorilla was discovered in 1902, this may just have been one explorer's attempt to create a new species by "splitting" it from the recognized lowland gorilla. As noted on the BergGorilla.org site, far more gorillas have been scientifically described than are currently recognized. Gorillas are variable, and unstable characters have often been used to distinguish purported species.

Wild Monkey Embraces, Bites 3-Year-Old Child in Illinois

Olive Branch, Ill., Oct. 21—(INS)—A long-tailed wild monkey raced out of the woods near Olive Branch, wrapped his arms around a 3-year-old girl and bit her before her father and a neighbor boy dashed a quarter-mile to the rescue.

The animal—nearly three feet tall and weighing 35 pounds—attacked little Mary Weichman yesterday and fled to a barn when her father, Clarence, heard her screams.

Weichman, who had no weapon, ran a quarter of a mile to a neighboring farm and returned with Teddy McKee, 13, who brought his rifle.

They found the ape-like creature hanging from a rafter by its tail. The youth wounded the animal with two shots. It fell to the floor and rushed at the boy before he could reload the gun but Teddy picked up a stick and clubbed it to death.

The little girl was only slightly injured.

Farmers in the area said they have been hunting what they thought was an oversized raccoon for eight years, puzzled by strange tracks and night noises in the woods at the rear of the Weichman home.

—Lima (OH) *News*, October 21, 1950.

Again, no reason within the text to suggest it is not an out-of-place primate, but it is interesting to note that the animal may have survived multiple Illinois winters. I'd suggest looking for local Illinois newspapers that might have photographed the animal to see if a species can be determined. Many primates were available for sale as pets in the 1940s-1950s.

Strange Mystery Animal Is Seen
in Trees Near Lemon Cove; It's Ape?

Lemon Cove, Calif., Oct. 30.—(U.P.)—The "strange mystery animal" which for three years has surprised and sometimes terrified residents in this area was seen again Saturday and speculation about its real identity started anew.

Ed Homer of Lemon Cove—a man whose word is accepted at face value by his fellow townspeople—vows he saw a monkey-like creature swinging in the trees in the foothills near here.

The only thing wrong with Homer's story, it was pointed out, was that monkeys (1) do not live in California mountains, (2) probably never have lived in California mountains, and (3) probably never will live in California mountains.

—Reno (NV) *State Journal*, October 31, 1938.

Another probable feral monkey, but note the newspaper's flawed reasoning. There are a number of exotic species introduced in California, and it isn't unlikely for certain primates to survive there, even if they don't establish a viable breeding population.

Reported Find Of Missing Link Will Be Probed
Sumatra Scene of Amazing Discovery
by Noted Dutch Explorer
Strange Creature Is Encountered On Trip
Emits Wailing Cry and Runs Swiftly Into Forest
at Sight of Man

By Charles A. Smith
International News Service Staff, Correspondent

London, Nov. 8.—That the "missing link" of the human race, the dream of all biologists, is alive, and that he exists in the Dutch East Indies, is the startling assertion of J. Van Herwaaden, prominent Dutch explorer.

Van Herwaaden declares that he encountered one of these creatures while hunting on Poeloe Rimau Island, situated in the southernmost part of Sumatra, last fall.

The description of the ape-man, as given by Van Herwaaden, so closely tallies with that which scientists would necessarily expect the ape man to look like that many learned scientific institutions have decided to send expeditions to the Dutch East Indies to try and find further evidences of the tribe. The government of the Dutch East Indies is also contemplating sending a party to the island with the same object.

Was Seeking Deer.

Van Herwaaden, in a graphic description of his encounter published in "Tropical Nature," states that he was traveling in October of 1923, in the Dutch East Indies, and visited the Island of Poeloe Rimau.

"Often early in the morning and sometimes during the day," Van Herwaaden writes, "I went out in search of a deer or a wild pig, without result, though I often saw tracks to encourage me.

"On one of these occasions, after having waited and watched for over an hour, I happened to notice something move in a not very tall tree, standing by itself in a small clearing.

"In order to obtain a good view of the object I went closer and walked around the tree. To my amazement I clearly saw a dark-haired being standing on a limb with the front part of its body leaning against the tree as if it wished thereby to hide itself and escape detection. This surely must be an ape-man, I thought. I was naturally very excited.

Refused To Budge.

"At first I only watched, with my rifle in readiness should it attack me. I then tried by calling and shouting to attract the creature's attention, but it refused to budge.

"I tried stamping against the tree trunk, but with no result. I put my rifle down and started to climb the tree, but when I was five or six feet above the ground I noticed the object move.

"It moved slightly away from the main stem, bent itself forward and sideways, so that I had occasion to observe its head, its forehead and its eyes, which looked into mine. Its movements at first were slow and deliberate, but after it had seen me it acted as though nervous and moved about quickly.

"I climbed down again in order to better observe the strange creature. The front of it was covered with hair, as well as its back, although the color of its front hair was lighter.

Hair On Head Dark.

"Its very dark head hair reached down below the shoulders, almost to the middle of the back, and was fairly well kept and straight. Its head seemed to be more pointed towards the top than that of a human. The brown-colored face was almost hairless, while the forehead was rather high.

"The arms reached to within a short distance from the knee when in a standing position. I did not have a good opportunity to see the feet, but the toes were well formed. The creature was about five feet in height.

"I heard, when I leveled my rifle at it, a wailing cry, 'Hoo Hoo,' which was immediately taken up by similar cries from the forest not far distant.

Dropped To Ground.

"I again laid down my rifle and climbed the tree, but when I had nearly reached the lowest branch the creature quickly ran to the end of the limb on which it had been standing, swung itself out and dropped to the ground from a height of about ten feet.

"The strange being, before I had an opportunity to reach the ground and grab my rifle, was easily ninety feet away from me, running for dear life and emitting sharp hissing sounds."

It is interesting to recall in view of Van Herwaaden's statement that the now world-famous skull of the pithecanthropus Erectus was discovered at Java, in the Dutch East Indies by Dr. Eugene Dubois about thirty years ago.

Scientists have been seeking traces of some such being as Van Herwaaden describes ever since the discovery of the fossils of the Neanderthal man.

—Reno (NV) *State Journal*, November 9, 1924.

This, of course, is a well-known early encounter with the cryptid ape, *sedapa* or *orang pendek*. Two points to note: First, the explorer's name should be Van Herwaarden. Second, the area in question (Poeloe Rimau) is better known as Pulau Rimau (translated, *Tiger Island*), north of Palembang, Sumatra. Apparently, the area Van Herwaarden visited was a forested delta of several rivers, including Pulau Rimau river, entering the Bangka Strait, rather than an actual island. It would not be the same as Pulau Rimau island, which is found off the southeastern tip of Penang.

Additional details (translated from Van Herwaarden's 1924 article in *De Tropische Natuur*) can be found in Sanderson's *Abominable Snowmen: Legend Come to Life* (1961).

Northeast Texas Mystery Animal
Draws Curiosity

Reno, Tex. (AP)—There must be something about Northeast Texas that brings out the mysterious and weird. Now it is something residents have named "the Critter."

Residents of this area just east of Paris have seen it—only at night—and their descriptions are enough to make anyone quake.

Those who have reported seeing it says it stands seven feet tall and has long, white hair over its body.

Bob Allison, 17, has seen it twice and fired two shotgun blasts in the critter's direction.

James Rogers of Paris, who runs more than 20 head of cattle on a small ranch east of Reno, has had to round them up twice because something stampeded them through the fence.

One of his cows turned up without ears—both had been literally chewed off—with large, deep gashes down her back. Only the head of the calf was found. The rest had been devoured.

"A pack of dogs, maybe. Or wolves," Rogers reasons cautiously, "but those slashes down the top of her back?"

His ranch is near where Wildcat Creek joins Cuthand Creek, east of the Allison home where the critter was seen on two successive nights.

The Allison youth kept noticing that the dogs and livestock were literally raising the roof night after night. He kept watch and spotted the thing.

The first time, it fled quickly.

The next night, Allison got within 30 feet and fired twice. "I think I may have hit it once," he says. "It sure left in a hurry."

Deputy Sheriff Arthur Fuqua checked and found nothing. Curiosity seekers have followed in droves.

—Port Arthur (TX) *News*, April 23, 1970.

White "Bigfoot" sightings crop up in certain parts of the country every so often. At this point, I'd suggest treating them no differently than any other Bigfoot sighting. The difference in coloration doesn't appear consistent enough to warrant a classificatory distinction. Explanations (inbreeding, color mutations, aging, environmental) are common, but irrelevant until a specimen is procured.

Strange Beast Reappears in Whiteside County

Sterling, Ill., June 30.—Sterling's wild beast made a sudden appearance the other afternoon near the home of Lyman Simpson, three miles southwest of Rock Falls. The animal, seen by a number of people, put a dog to rout and was pursued by several men with guns, but made its escape through a field of oats.

The animal was first sighted by several small children playing in the road close to the Simpson farm. They ran home, badly frightened.

While the men were seeking weapons, William Monnier of Morrison, who was a guest at the Simpson farm, and Clarence Simpson, both boys of 14, leaped on their wheels and raced down the road.

The dog sighted the animal first, turned tail and beat a hasty retreat. Both boys saw it, turned and hastily rode back to the Simpson home.

They described it as larger than an ordinary mastiff dog with white shoulders and yellow spots on a tan background.

The night before the animal killed a calf belonging to a neighbor of Mr. Simpson. It was seen at that time, but by the time pursuit was arranged it had gotten away.

—Freeport (IL) *Journal-Standard*, June 30, 1938.

"Beast" sightings, identities always in question, are very common in newspapers during the mid-1900s. Most "varmints" were hunted ceaselessly in agricultural communities, so that encounters with native canines and felines eventually became news items in and of themselves. The natural variation in those species, and the spread of coyotes and feral dogs, created an environment where even rural individuals would not necessarily have been capable of accurate identification of even known species. So, most of these reports are best taken with a generous helping of salt.

Here is another of this sort:

Fear Peculiar Animal
Men and Dogs of Minnesota Are Terrorized
By Ferocious Wild Beast.

Winona, Minn., Nov. 8.—Roaming in the wooded land where it evidently has a hidden place in which to hibernate, is a large, strange animal, so ferocious that it has caused men ordinarily frightened at nothing to flee in great fear at the sight of the beast, according to advices from Pickwick at the lower end of Winona county.

Reports of seeing the beast have persisted for several weeks. What is is none who have seen it will say.

The most reliable information thus far is said to have been gained from Carl Nelson, a farmer; residing on the edge of the infested woods. Nelson swears he saw the beast plainly and that it was light grey in color, striped and about as large as a yearling calf.

David Hoffers a retired merchant went to the woods with two good hunting dogs and a high powered rifle. Several miles below Pickwick his dogs picked up a trail. They followed it to a heavily wooded place which backs into a rocky draw. The dogs began to bay, then suddenly broke and fled to their master, tails between their legs. Huffors turned around and went home. He said he didn't see the animal—didn't even have a desire to see it. The fear of the dogs satisfied him, he said.

Farmers around Pickwick believe the animal escaped from a circus, has worked its way up the Mississippi river and is unable to cross it.

Those who have sent dogs on the trail of the beast declare they become greatly excited when the trail is first picked up, but after following it for some distance, break for home, displaying unusual fear.

—Lincoln (NE) *Evening Star Journal and Daily
News*, November 8, 1919.

If the description is accurate, it's a fascinating story. But, accuracy cannot be readily judged from a single sighting report. This is where a focused search through the local Minnesota newspapers from this period might prove useful in further evaluation.

Not to be excluded from discussion of mystery beasts is the possibility of exotic escapees. While it may be used as a catch-all excuse, anyone with experience with exotics knows that they do sometimes get loose, and not all are immediately recaptured. We've heard many stories of "maned" lions and black panthers in North America, but there are also stories of striped cats and spotted felines. They may not get as much publicity, but they are present, and escaped exotics are as good an explanation as misidentification or relict North American tigers. (As noted in *Cryptozoology: Science and Speculation*, the North American fossil cat *Panthera atrox* was more likely a tiger than a lion—and highly unlikely to still exist.) Bearing that in mind, here are a couple of "tiger" stories:

It Must Be A Tiger
A Wild Beast Which Is
Terrorizing McDonough, Ills.

Lewiston, Ills., June 8—The piteous screams of a horse in the stable brought Frank Chatterton, a farmer of Bernadotte, to the scene Sunday afternoon. As he entered the lot a huge beast sprang from the stable door and, after bounding into an adjacent field, crouched low, uttering deep growls, while its long tail waved slowly to and fro. Chatterton was horrified, and, fearing either to advance or retreat, gazed helplessly at the big beast, which in a few moments slunk away into a patch of timber.

Feeding on a Steer's Carcass.

The horse lay on the stable floor weltering in a pool of blood which flowed from a dozen wounds. The animal's sufferings were soon ended by a ball from a rifle. Chatterton declares that the brute he saw is not a panther and that he believed it to be a tiger. The excitement is augmented by the report that John Hulvey, residing some miles from here, came across a large animal in his field Thursday, which was feeding on the carcass of a steer. The animal was disposed to show fight, and Hulvey retreated in haste. The animal's screams have been heard and its huge tracks have been found in the Spoon river bottoms.

Terrorized the County.

For the last three months McDonough county has been terrorized by this creature. A party of hunters surrounded the brute in the Crooked creek bottoms a few weeks ago, but the dogs would not attack it. The men caught a glimpse of the animal, and were so frightened that they gave up the chase. About three years ago a menagerie, while crossing the Crooked creek bottoms, was caught in a storm. A cage containing a tiger was overturned, and the tiger escaped. This is believed to be the animal which is now terrorizing this section.

—Decatur (IL) *Morning Review*, June 9, 1891.

Jungle Beast at Large in State
Tiger Reported Roaming the Fields
in the Vicinity of Norfolk.

Norfolk, Neb., Sept. 27.—The tiger which has been reported seen in several nearby counties during the last few months, and last week was believed to be on a farm near Newman Grove, was seen standing on the banks of the Elkhorn river between here and Battle Creek yesterday, according to a report brought to the city by Mr. and Mrs. Ed Muffly of Battle Creek. They declare they got a good view of the animal as they crossed the bridge over the river. Farmers have been warned as it is feared the beast will make raids on livestock.

It is supposed the tiger escaped from a circus last spring.

—Lincoln (NE) *Star*, September 27, 1923

Here's a "spotted" feline story, almost certainly cougar, but with more than one possible explanation:

Hunt Spotted Lion

Ashford, Wash.—A mountain lion curiously marked with black and gray blotches on his tawny hide has been seen by three woodsmen near here at various times this winter. The spotted cat is being hunted to death in the big drive government hunters are making in Mount Ranier National park. It is believed the cougar was caught in a forest fire and severely burned in spots, and the hair is returning black and gray where the tawny skin was wounded.

—Placerville (CA) *Mountain Democrat*, February 16, 1924.

Although rare, there are a few photos known of adult cougars retaining spots, possibly a form of paedomorphism.

Ogopogo, Loch Ness Monster, Now Bunyip
Australian Swamp Creature Said Just Shy Man-Eater

By Gordon Tait

Sydney, Oct. 3.—(AP)—The bunyip, a fearsome swamp creature that is to Australia what the Loch Ness monster is to Scotland and "Ogopogo" is to British Columbia, has come into the news again.

Ern Elphick, of Gundagal, southeastern New South Wales, told a reporter he had seen "a thing that might be a bunyip" on a lagoon near Gundagal, fired four shots at it from a shotgun, but the "thing" submerged, and he didn't see it again.

Every year or two someone reports he has seen a bunyip in a lonely swamp or shallow lake, but no one has yet killed or captured one.

The bunyip, according to legend, is a shy man-eater, with a particular liking for aborigines, who feared it before the white man came to Australia to live 160 years ago. But beyond its eating habits, there is no generally accepted idea of what the bunyip looks like.

Elphick said the one he saw when he was accompanied by a friend had a body a little more than three feet long, apparently covered with hair. He said there was a neck like the thick part of a swan's neck, no head, but two ears hanging from the upper end of the neck. The creature, Elphick told a reporter, "makes a noise like the moaning of a bull and spouts water eight feet into the air."

—Lethbridge (Alberta) *Herald*, October 3, 1947.

There is a follow-up article to this news account, where someone shoots a musk duck and suggests it was responsible for the sighting. I'm a bit skeptical of that explanation. Still,

some of the better possible candidates for bunyips appear to be
pinnipeds rather than carnivorous man-eating unknowns.

Mysterious Beast Is Spreading Terror
in Tennessee Community

South Pittsburgh, Tenn., Jan. 16—(AP)—A
mysterious animal as "fast as lightning" and like
a "giant kangaroo" is spreading terror through the
Hamburg community.

The creature first appeared Saturday night. It
killed and partially devoured several German police dogs. The next night it killed other dogs and
some geese and ducks.

Farmers are carrying their shotguns to the
fields for fear of the beast and others are going
about their daily work armed with pistols.

Rev. W. J. Hancock, negro minister, saw the
animal. He said "it was as fast as lightning and
looked like a giant kangaroo running and leaping
across the field." It had just killed a large police
dog and had left nothing but the head and shoulders of the victim in the owner's yard. Frank Cobb
also saw the thing. He said it was unlike anything
he had ever seen but that it resembled a kangaroo.

A searching party tracked the animal up a
mountainside to where the trail disappeared near
a cave.

—Lincoln (NE) *Star*, January 16, 1934.

This is one of the reports primarily responsible for the "Devil
Monkey" classification in cryptozoology. Personally, I think
there's a good deal more to be investigated concerning a bounding,
leaping mid-sized creature reported from various parts of the
Appalachians and surrounding forests. This is where a few local

investigators could do some serious work on a barely-discussed cryptid. I know one contributor to this anthology has recently spoken with witnesses of a similar, though perhaps more arboreal, creature. Hopefully, more investigators will follow suit.

Explorer Attacked by Prehistoric Bird

London.—Baron Munchausen, says the Daily Express, is reincarnated in the person of Ivan Levey, who describes an encounter with a prehistoric moa in the wilds of North Island, New Zealand.

He was assaulted, says Levey, by a 14-foot moa, a class of beast that had been generally supposed to be extinct.

Its color was light brown and its body huge and bulky.

There were no signs of wings, the legs were disproportionately massive, almost elephantine, and the three-toed feet were "simply ponderous."

A small head rested on a long ostrich-like neck. The moa uttered a deep, booming noise, says Levey.

—Stevens Point (WI) *Daily Journal*, September 4, 1919.

I just wanted to point out here that while there are several books on Australian cryptozoology, they are much rarer on New Zealand cryptozoology—which is a shame. It really would be nice to see a New Zealand investigator put together a comprehensive examination of the mystery animals of that country.

Strange Sea Fish.
Ship Sails Through Miles of Funny-Looking Things.

Victoria, B. C., special San Francisco *Examiner*. Capt. Amesbury, of the British ship Puritan, presented a remarkable specimen of ocean fauna to the British Columbia Museum. It is a strange barnacle-like mass of heads, like half-closed tulips, folding over bunches of stringy, clammy feelers, linked by a short cylindrical neck to a pulpy body. It is somewhat like a goose barnacle, but the resemblance is not very great. Shipping men to whom the strange fish, animal, or whatever it is, was shown, have never seen in all their experiences such a creature. The Puritan passed through miles and miles of them while scudding along many hundred miles out from the Oregon coast a week ago. Capt. Amesbury at first thought they were seaweed. The sea was running high, and some of it was swept up on the fore-castle. It was then that Capt. Amesbury saw that, instead of inanimate seaweed, he had come across a strange creature of the seas. One of these washed on the ship he fed with oatmeal, on which it waxed fat, until about two or three days from port it died and was placed in alcohol. After the ship had cleared the thousands of acres of the unique creatures she passed through many acres of Portuguese men-of-war, as sailors name the ocean jelly fish. Capt. Amesbury and others are of the opinion that the strange creatures were cast up from the ocean bed by the recent earthquake shock.

—Mexia (TX) *Evening Ledger*, October 5, 1899.

Well, there are all kinds of oddball marine invertebrates. No surprise there. The next story is a bit more interesting:

Roared Like A Lion.
A Coaster Captain's Story Of A Remarkable Whale.
It Didn't "Blow," but Its Roar Was Something

Awful—It Was Eighty Feet In Length and
Had a Queer Looking Head. Tried to
Swamp the Schooner.

"No," said Captain J. A. Crossman of South
Portland, "we didn't encounter the sea serpent,
but we had a strange experience with a whale, and
I don't believe anybody ever had the like before.
I've been at sea, man and boy, since I was 9 years
old, and I never saw the like of the whale we encountered. I never saw a whale before that didn't
blow, but the one we met didn't, but it gave a roar
that was awful."

"It sounded like the lions in Central park, New
York," said Miss Houston, who was one of the
party that had the strange encounter.

"It certainly was more like the roar of a lion
than anything else I can think of," said Captain
Crossman.

The schooner Grace Webster, Captain
Crossman, was on here way from New York to
Portland with 414 tons of coal. Besides the captain
and crew Mrs. Crossman, her daughter and Miss
Houston were on board.

The schooner was about ten miles off Wood
island and making good headway under full sail,
the mate, Merrill Crossman, at the wheel, when
there was a sudden commotion ahead, and the
great head shot up into the air and was on a level
with the deck.

One of the crew first sighted the strange creature and called Captain Crossman, and in a moment
all on board but the man at the wheel were looking at the strange sight. They saw before them an
enormous head, one mass of great bunches,
through which the wicked looking eyes of the creature gleamed. They expected the whale, if such,

would "blow," but it did not then or after. Once a narrow thread of what looked like steam shot up, but not a drop of water was sent into the air. As they looked at the creature it roared savagely, and then drew close up to the side of the schooner, giving them ample time to observe the head, and all agree that it was very broad; that it tapered almost to a point, and that it was not very thick through the thickest part. The creature was about 75 to 80 feet in length and had a very broad tail, very different from that of an ordinary whale. In fact, at the time there were three or four whales in sight, and they had no difficulty in noting the points of difference between them and the stranger.

The great creature went down head first and then made a series of attempts to strike the side of the schooner with its tail. It did not succeed and swam around them, roaring loudly, in evident anger. Then it went down and under the schooner.

Captain Crossman, who had watched for this movement, gave orders to be ready to lower the boat, fearing that the whale might come up under them and break them in two. It was very fortunate that they were not forced to lower their boat, as it proved later to be leaking and would not have carried half their number safely to land.

For more than an hour the whale continued its remarkable acrobatic performance, standing on its head, with its tail waving in the air most of the time. It seemed bent on hitting the schooner, and it took the best of good seamanship to prevent an encounter.

At last the whale seemed to get tired of what had been fun at first, and it headed for the westward.

Captain Crossman is of the opinion that this strange whale must have been mistaken for a sea serpent many times. Seen but a short distance off, the head would look more like that of a great serpent than of a whale.

The creature, while making its long and repeated attempts to hit the schooner, continued its roaring when above water. It would scrape against the side of the schooner and then would draw off, seem to be calculating the distance, and then strike. The schooner was kept off at the right moment, and the creature missed the vessel every time. It was an odd experience, and for a time there was something closely resembling a panic, the women being badly frightened.

Miss Houston said that the sight of the great mass standing almost upright in midocean was something not to be forgotten. Captain Crossman is uncertain whether the whale is a natural fighter or whether it was frightened when it came up out of the water, and as a result of its confusion made the repeated attempts to sink the schooner. One of the crew hit it with a bolt, and many times Captain Crossman said he could have hit it with a board from the deck. He did not venture to do anything to further arouse its anger and let it go in peace.—Portland (Me.) *Press*.

—Indiana (PA) *Progress*, January 1, 1896.

Of course, what is particularly interesting is the implication that a cetacean might be responsible for sea serpent reports. Of course, this presents its own difficulties, but at least we have a track record of new species of whales and dolphins being discovered within the last century.

Freak of Nature.
Putnam County Trapper Captures Two
Freak Specimens of the Fresh Water Turtle.

One of a pair of soft shelled fresh water turtles that are either remarkable freaks of nature or a new species was captured at Ottawa by Emory Stambaugh, a veteran hunter, trapper and fisher. The tortoise had a curious hump on its back about the size of a large orange. The shell, which was about eight inches in diameter, extended about two inches on each side of the hump and was fairly regular in shape, having a general appearance of a narrow brimmed, high crowned Mexican hat. The spinal column curved up along the center of the hump. The under shell of the turtle was of normal shape, but was milk white in color, and without markings, instead of being yellow with dark markings, as is usual with river turtles. Stambaugh said that several weeks ago he shot a similar turtle, which he then thought had got its back out of shape by being caught under a heavy stone during its winter hibernation. He had with him the shell of this other turtle. He had eaten the meat and said that the hump portion was unusually sweet and fine flavored. When he found the second hump-backed turtle he decided that two individuals could hardly have sustained such precisely similar accidents. As the humps are very symmetrical and the ribs and vertebrae inside show no distortion, it looks as though a hitherto unknown species had been discovered. The living turtle will be sent to the state university for examination. It has been suggested that the new species, if it proves to be such, be called the dromedary turtle.

—Van Wert (OH) *Daily Bulletin*, June 29, 1907.

While it might be too coincidental for two similar accidents resulting in a humped-shell, there are other possible shared conditions: genetic anomaly, parasitic infection, or a metabolic bone disease. It would be interesting to know if the Ohio State biological museum (or one of the other state universities) has a record of this turtle.

Finds "Water Elephant."
African Explorer Discovers Strange Beast of Unknown Species.

Special Cable to *The Washington Post.*
London, Feb. 5.—Rumors of the discovery of a hitherto unknown animal in central Africa by Dr. Trovessant has prompted the eminent naturalist, Richard Lydekker, to write to Dr. Trovessant, who sent him a copy of the Paris scientific journal Nature, in which he writes:
"We just obtain additional information of a mysterious animal inhabiting the lakes of central Africa. The natives call it a water elephant on account of its aquatic habits. Lepetit, one of the explorers sent by the Paris Museum of Natural History, informs me he is at Tombamayl, on the northern shore of Lake Leopold, in the Belgian Congo. Five of these animals halted at a distance of 500 yards from Lepetit, enabling him to observe them for some seconds. The trunks and ears are remarkably short.
"The height does not exceed above 6 feet. There were signs of tusks. The animals' foot prints in the mud are different from an elephant's. The animals, catching sight of the traveler, plunged into the water, leaving only the summits of their heads and their trunks exposed."

—Washington (D.C.) *Post*, February 6, 1911.

This would appear to be part of the water elephant/pigmy elephant back story. Note, however, the following article, which appears to merge the water elephant with accounts of another mystery beast to suggest a distinctly different candidate:

Water-Elephants; Formidable Beasts Seen by Few Whites
French Collector Reports Having Seen
the Mysterious Animal—Smaller Than Elephants
With Longer Necks—Dangerous to Boats.

When J. D. Hamlyn returned in 1905 from his successful expedition to the Congo in search of living chimpanzees, one of the most interesting stories he had to tell me, says a writer in the London Saturday Review, was concerned with the belief of the local natives in a large and formidable aquatic beast which no white man had ever seen. Recently this report has received very strong confirmation, for a French collector, M. le Petit, whom Mr. Hamlyn had himself met while in the Congo region, has since actually seen the mysterious animal. From his account, and from Mr. Hamlyn's comments on it, we can get a pretty good idea of the creature. M. le Petit saw five specimens, and watched them both on land and in the water, though at a distance of more than a quarter of a mile; he made out, however, that they were much smaller than elephants, not standing more than six feet high. They had no visible tusks, and their ears and trunks were short; the neck, on the other hand, was longer than in the elephant. Mr. Hamlyn adds, in commenting on this account in the Star, that the creatures were said to be hairy, with very thick hides; to be dangerous to boats, and capable of staying under water for some time.

They did not, he was told, associate with either the hippopotamus or the elephant.

This account seems to suggest some kind of tapir, a beast once reported from Africa. Tapirs take to water freely, although not exactly aquatic animals, nor are they known to be ferocious. Moreover, all the few species of tapir known are American, except one outlying form in Malaysia, and none is anything like six feet in height. On the other hand, West Africa is just the place in which a new tapir might be expected. Fossil remains of tapirs are known from European deposits, and there actually exists in West Africa a little beast which belongs to a genus first described from a European fossil. This is one of the small primitive hornless ruminants known as chevrotains or mouse deer, and all its relatives live in southeastern Asia, some of them alongside the Malayan tapir. A tapir, therefore, might very well occur in West Africa as well as a chevrotain: more than that, it might very well be large in size and fonder of water than tapirs as we know them, for the West African chevrotain is the largest of the small group of species to which it belongs and is more aquatic than its relatives; in fact, it is the only one of them which is aquatic in habit. If the suggested new tapir is really a large beast, it may very well be dangerous, especially to ill-armed men in canoes, for the existing tapirs, though pacific, are very strong animals, and can bite severely.

Whatever the African water elephant may ultimately turn out to be, we have, I think, a chance of actually seeing the legendary water elephant of India, now that a specimen of the great elephant seal, or sea elephant, has been for the first time received at the zoo. This individual does not look at all like an elephant at present, it must be admitted;

its form is very much more like a fish. It is not bigger than our common seal, and all one notices about its nose is that it can be wrinkled up in a comical manner: there is no approach to a trunk. But the adult male of this seal is really a rival to the elephant, among sea beasts; it is larger even than the walrus, reaching five yards in length, and it has a trunk, though only about as long as a tapir's. Now it is a very curious thing that while Sanskrit writers speak of a water elephant, a curious beast is depicted in ancient Indian sculptures which shows a combination of elephant and fish, just as the legendary mermen and mermaids of Western myth combine the fish and man. It has been suggested that the origin of this supposedly mythical beast was a hippopotamus, known by fossil remains to have once existed in India; and the teeth given to it in the representations are certainly not in the least like the tusks of an elephant. It has, in fact, typical canines directed downward in the upper and upward in the lower jaws. It is pretty evident, therefore, that the old sculptors knew that their beast was not like an elephant in the mouth. It will be noticed, too, in a sculptured figure of it on one of the stones of the Amravat tope in the British museum (in a top corner on the landing) that the trunk is short and that no ears are visible.

The teeth, however, are not of the exaggerated size which characterizes the hippopotamus's huge canine tusks, and would suit the sea elephant much better, and the short trunk and fishlike body make the identification with this creature quite reasonable. It may be properly objected here that the sea elephant is essentially a southern animal and that neither it, nor any other seal for that matter, inhabits Indian waters. But, on the other hand, a local race of the sea elephant existed till the last few years on the coast of California, in a corresponding latitude in the opposite hemisphere; and there seems no reason why stray individuals might not have ranged to the north on the eastern side. Familiar the creature could not have been; the conventionality and inaccuracy of the representation, in sculptures where well known animals are faithfully enough depicted, is evidence of that. Curiously enough there is evidence that a bird companion of the sea elephant may also stray into Indian waters: the little diving petrel, a true denizen of the southern ocean, has been seen there both by Sundevall, the Danish naturalist, nearly a century ago, and in recent years by myself. Diving sea fowl, it may be mentioned, are almost as unlikely to be seen in Southeastern Asiatic seas as seals are: there are no auks, penguins or divers there, and the cormorants keep mostly to the fresh water.

—Syracuse (NY) *Herald*, August 13, 1911.

A few years later, and nothing more is known:

Secrets of Africa
Rumors of "Yet Another Strange and Unknown Beast" in Kongo Region.

From the Pall Mall *Gazette*.

Ex Africa semper aliquid novt. The proverb of the ancient world still holds good in the bustling days and amid the unflagging activities of the twentieth century. The lastest report from what used to be known as the Dark Continent appears in a provincial contemporary, the London correspondent of which records a report of the discovery of "yet another strange and unknown beast" in East Central Africa.

Particulars are said to have reached the Natural History Museum of the existence of an animal "about the size of a bear, tawny color, with very shaggy long hair." This interesting creature is also described as "short and thick-set in the body, with high withers and a short neck and stumpy nose," and "its existence is vouched for by more than one official."

Inquiry at the museum by a representative of the Pall Mall Gazette fails, however, to confirm these attractive details. In the eyes of the authorities at South Kensington "particulars" of such discoveries have to be of a definite and material kind in the shape of some portion of the animal, bones, for example, or a piece of skin.

Nothing of that sort is at present forthcoming. "Rumors have, however, reached the museum," said an official of the mammal department "of the existence of a hitherto unknown animal, possibly such as is described, but we have no 'particulars,' and can, therefore, substantiate no account of details. When the okapi was discovered 'particulars' were forthcoming in the form of belts made from its skin and worn by the natives."

"You do not, then, credit the account?"

"We don't say such an animal does not exist. All that can be said is that no 'particulars' are to hand up to now. There was a rumor some time ago of the discovery of a water elephant which, apparently, was of the nature of a very substantial tapir, but nothing seems to have come of it. If we could have a proper systematic survey of Central Africa, it would probably result in the discovery of any number of new creatures.

"The Kongo region, whence this rumor in all probability arises, is almost unknown from a zoological point of view. There are, for instance, any number of new monkeys there, concerning which we find nothing in the books of ten years ago. A few men are engaged in natural history research on the East Africa districts have been swept by the big game expeditions of Roosevelt and others. But without doubt there are many unknown animals yet to be discovered in the Kongo region."

—Washington (D. C.) *Post*, October 13, 1913.

Eventually, the tapir theory fades away, and the focus is back on these small elephants as a pigmy species.

Pigmy Elephant Discovered in Congo

London, March 2.—A species of elephant said to be hitherto unknown, a real pigmy, has just come to light in the Congo, and two stuffed specimens have been sent to England, one for the Natural History museum in London and the other for a similar institution in the United States. The two animals are both adult specimens, as shown by the teeth. Their height is only about five feet six inches, and the tusks of the female weight only about two pounds a pair, as against 200 to 220 pounds in a well grown African bull elephant. The legs, ears and tail are of distinctive character. The species is known to the African natives as the swimming or water elephant, making its home in marshy country.

—Fort Wayne (IN) *Journal-Gazette*, March 3, 1918.

The pygmy elephant saga was discussed in the late Dr. Bernard Heuvelmans' writings, and is summarized well in both

Eberhart's *Mysterious Creatures* and Newton's *Encyclopedia of Cryptozoology*. This isn't the only strange elephant reported:

Betrayed By Their Long Hair.
Animals Which Claim a Descent From the Supposed Extinct Mammoths.

Among the distinguished passengers of the North German Lloyd steamship Werra, which arrived at her dock in Hoboken yesterday morning, were two animals which had attracted a good deal of attention from the passengers during the voyage, and which Charles Reiche, the dealer in rare animals and birds, who imported them, claims to be specimens of the Elaphas primigenus, or mammoth, an animal which naturalists affirm has been extinct for centuries. The animals were confined in two stalls, in the lower hold of the steamer, and throughout the voyage they behaved very well, neither of them suffering in the least from seasickness, and both eating their meals with a relish presumably created by the salt sea air. The lady passengers on the Werra made pets of them and visited them daily, and, in addition to the boiled rice and milk, mixed with bran, the hay and the oats furnished them by their keeper, they received numerous delicacies from the cabin table. They were as gentle as kittens, and repaid the kindness of their benefactors by allowing them to pat their hairy heads and extending their slender trunks to shake the hand of the fair providers of their extra lunches.

A number of gentlemen visited the Werra to see the animals disembarked yesterday, on the invitation of Mr. Reiche. When they were taken from their stalls and marched out on the lower deck of the steamer to be raised to the dock a fine opportunity to examine them was offered, and as they were as tame as a pet dog and allowed themselves to be pulled about and turned around at the pleasure of the visitors every peculiarity of the animals was easily noted. The larger of the two, which is named Phunga, is about 4 1/2 feet high, about 4 feet long, and her body is thick and massive, measuring about 3 1/2 feet in circumference. The smaller, Quedah, is about 2 1/2 feet high, about the same length, and broad in proportion. Both bear a great resemblance to the ordinary elephant, but the head of each is somewhat more elongated than that of the elephant, and the ears, which are thick and hard to the touch, are set back firmly against the side of the head, instead of lopping over, like the ears of the common elephant. The skin is hard and tough, of a dark slate color, and on the larger of the two animals it is well covered by a growth of stiff black hair, resembling the bristles of a hog, but longer, reaching a length of from 6 to 8 inches. This hair is longer and thicker about the neck and around the buttocks than on the rest of the body, and on the legs and trunk it is missing altogether. The legs are shorter than those of the ordinary elephant of the same size, and the trunk is small at the mouth and tapers gradually to a very delicate circumference. The tail is long and almost touches the ground as the animal walks. The smaller of the two, which Mr. Reiche believes to be a dwarf, possesses all the peculiarities of the larger, with the exception that the hair or bristles is not so thick. It is believed that the larger of the two beasts is about 10 years old, and the smaller 6 years.

The fact upon which Mr. Reiche bases his belief that these two animals are specimens of the mammoth, or Elephas primigenus, is the existence of the hair on their bodies. The only perfect remains of a mammoth found in recent times were those discovered by Schumachoff, in 1799, on the shores of Lake Oncoul. It was frozen fast in the ice, but four years later, when the ice melted, Schumachoff, who was a hunter, visited the spot and secured the tusks of the animal. In 1806 Mr. Adams, of the St. Petersburg Academy, secured the animal, and a large quantity of hair was then attached to its skin. The hair was its protection against the cold climate of Siberia. The skeleton of this animal is now preserved at St. Petersburg. The two living elephants which Mr. Reiche believes to be descendents of the same species were captured on the Malay Peninsula about six months ago, and purchased from the Indian who trapped them. They will remain at Mr. Reiche's stables at Tenth and Hudson-streets, Hoboken, until tomorrow, when the larger of the two will be shipped to Philadelphia, where she has been rented for $1,500 a month to a museum. Mr. Reiche values the two animals at $50,000.

—New York (NY) *Times*, September 21, 1884.

Mammoths? Well, no...
Juvenile Asian elephants are covered with reddish-brown bristly hair. Sounds like Mr. Reiche was pulling a fast one. Charles Reiche was a recognized animal dealer (and eventually started up an aquarium in New York with a partner), so he should have been able to distinguish young Asian elephants.

But, jumping back to Congo mystery animals, the next news article should sound familiar:

"Njago Gunda," Native of Africa
Will Roosevelt Find Strange Animal?
Prof. R. L. Garner, the Famous African Traveler,
Describes the Strange Beast That Has Terrified the
Natives in the Dark Continent.

(Written Especially for the New York *World* by
R. L. Garner.)

Far away in the depths of the great forest of
Africa there are, no doubt, many strange forms of
life, both animal and vegetable, that are yet unknown to the outside world. From time to time
new types of both kingdoms are discovered and
made known to science; but that vast extent and
wild character of the regions yet unexplored justify
the hope that future discovery may add many new
links to the chain of life.

Only a short distance south of the equator and
not far from the western seacoast of Africa is a
large lake commonly known to geography as Lake
Fernan Vaz, but in the native tongue is called
Ellwa Nkami. It is fed by several streams flowing
from the southern branch of the Ogowe river, and
has its outlet at the bar of Fernan Vaz.

In its general outline Eliwa Nkami somewhat
resembles a Gothic L. Its total length is some fifty
miles, and its width varies from a mile or so to
eight or nine.

In the midst of all this wealth of beauty the
great lake spreads out its amber waters over many a
dead man's bones, and unseen danger lurks in every
nook. This charming lake is one of the favorite haunts
of fierce amphibians. In many places schools of
hippopotami may be seen at any hour of the day.
Those who frequent the lake in their canoes those
places are well known and easily avoided, but those
ugly beasts roam about under the surface of the

water and are often encountered at most unexpected places. With little or no provocation they often attack and sometimes destroy canoes and all their occupants.

Crocodiles also infest these waters, and while they do not attack canoes, they are able at any time to attack a man who gets into the water. There are also sharks of the smaller kind and a variety of other fishes that will attack anything that comes within their domain.

The people who occupy the basin of this lake are known as the Nkamis. They are not one of the largest tribes of Africa, but are a people of average native intelligence, and perhaps more than average cleverness in adopting the manners of the few white people who come among them. They were formerly one of the chief slave-trading tribes of the coast, and many of the black people of America are descendants of them or of contiguous tribes upon whom they once carried on perpetual war in order to capture prisoners to sell to the white slavers who came to this section to buy them. Many traces of the old stockades are still visible along the coast.

As far back into the past of the memory of the oldest men and women of this tribe can go, there have been, and still are, current among the people authentic accounts of a strange monster of gigantic proportions and of great ferocity inhabiting the deep waters and dismal marshes of the lake. It has been frequently and recently seen by many living witnesses, and many deaths of natives caused by it testifying to its reality. Numbers of canoes have been smashed into splinters by it, and scores of human victims are charged to it. For some months in succession it has loitered about the mouth of Mpivie creek, a deep inlet of the lake, where it attacked or chased every canoe that came near its haunts, resulting in the total ruin of many cones and the cost of more than twenty lives.

In the native language the animal in question is called njago gunda. The former term is the vernacular for elephant in its generic sense, and is applied as such to the two distinct species that inhabit the forests and plains of this region and both of which are well known to the native hunters. The latter term, gunda, is the specific name of that part of the exoskeleton which develops in the form of coarse, wire-like bristles, of horn-like texture, which grows at the end of the tail of all elephants. In its use in this instance, however, it is not only intended to describe these as a marked characteristic of the animal, but it is meant to imply something of his frightful appearance by reason of the unusual abundance and great size of them and their peculiar distribution.

All accounts agree that njago gunda is more than twice as large as an adult male elephant of the ordinary kind. It is somewhat darker in color and much more active and rapid in its movements. Its proboscis is only a little more than half the length of its head, and near the base of it is an oval hole on each side resembling nostrils. Their functions are not positively known, but they are believed to be valves through water is drawn into the nasal cavity and ejected through the proboscis. The ears are comparatively small and pointed. Above each eye and obliquely above each nostril or valve is an enormous fan-shaped tuft of those stiff bristles, or ngunda, more than a foot in length, the longest of them possibly eighteen inches. Along each side of the proboscis is a thickly set row of them, somewhat shorter than those elsewhere about the face. From the upper edge and point of each ear projects an array of them and from the frontal prominence along the sagital ridge and as far as has been seen along the spine is a dense row of them even much larger than those about the face.

Njago gunda is not only a creature of horrid aspect: he is quite as ferocious as he looks to be, and is very justly regarded by the natives as the chief terror of the lake. Without the slightest provocation he fiercely attacks anything that approaches him. In so doing he suddenly erects those tufts and rows of hard black bristles, rears his head and proboscis above the surface of the water, and with a long piercing shriek rushes like a comet upon his helpless prey. When he comes within eight or ten yards of the object of his attack he jets from his short, thick proboscis a terrific stream of water with the force of a fire-plug. In this manner he instantly capsizes or swamps a cone and stuns or stifles its occupants. Without pausing for an instant he presses the attack, and with rapid strokes of his deadly proboscis he crushes everything in reach.

The volume of water thrown by njago gunda is said to be as thick as the arm of a man and sent with force enough to kill a man at a distance of four or five yards. In the native language it is called "mbumba nengo," meaning rainbow of water. The bristles act in concert, and during the assault the fiend causes them to rise and fall with rapidity, furiously lashing his hard skin and causing a weird, swishing sound which chills the courage of the bravest of men.

Not long since a convoy of four canoes coming down the Rembo Nkami was attacked by one of these monsters. Every canoe was smashed into fragments, and nine out of fourteen people in them have never since been seen or heard of. The catastrophe occurred near a bank of the river, and five men thereby escaped and made their way by separate routes through the bush to a native village. Stray bits of wrecked canoes were afterward found along the sides of the river, but no trace of the nine victims, of whom two were women, was ever found.

In all essential points the accounts given by the survivors sufficiently agree, but the minor details are not enough like to warrant the suspicion of a concocted story and the separate versions of this incident, as give by those five men, not only coincide with each other, but the fundamental facts concur with the reports of the few survivors and witnesses of similar attacks of the brute made elsewhere.

Some years ago one of these animals was found dead in a deep morass near the mouth of the Rembo Nkami. No one knew the cause or time of its death. It was only discovered by the smell, and no native would venture near it. It lay there and decayed, and the bones are possibly there to this day; but no native of this country could be induced to search for them or to touch one of them, if found, because the animal is regarded as mbuiri or fetish. There are plenty of people now living the vicinity who ventured near enough to get a partial view of the carcass, but not one of them will go near the spot where it perished, and the price of a kingdom would not induce any one to dig in the marsh for the skeleton.

The general belief, however, among those who know most about the animals is that they are exceedingly rare, and only a very few at most have ever been seen in the lake of its environs. It is definitely known that four or five different ones have been seen, only one of which is thought to be a female. Not one has ever appeared during the dry season, and it is very exceptional that one is seen when it is not actually raining.

During a period of more than twelve years and a resident of more than two years among the Nkamis I have made the most searching inquiries that I am capable of, and I am finally persuaded that there is a foundation of truth upon which the story of the njago gunda rests. The matter-of-fact manner in which accounts of the beast are narrated and received, and all these separate witnesses, without rehearsal or previous consultation, concur in the main facts both as to the appearance and conduct of the monster. It has been seen by more than two scores of peoples at the same time, and at different times and places by many hundreds, all of whom give similar, though not identical, accounts of it.

It may be much smaller in size and much less ferocious than it has been described to be, but that it is some rare beast of peculiar form and frightful aspect there is little or no reason to doubt. Whether it ultimately proves to be some new species of pachyderm, as it is now believed to be, some strange cretacian, as it is more likely to be, or some reptilian novelty, as it is possible to be, cannot be determined from the meagre data now available.

—Fort Wayne (IN) *Journal-Gazette*, June 13, 1909.

This creature, spines and all, is part of the folklore tradition that includes the more popularly written-about Mokele-mbembe. Expeditions to these African waterways continue; that is about the only realistic investigative methodology that has any chance of answering questions about the creature's true identity.

New Kind Of Bear Is Discovered In Oregon
Resembles Small Grizzly,
United States Biological Expert Announces

The discovery of a new species of bear was announced at a meeting of several hundred Department of Agriculture employees with Henry C. Wallace, secretary of agriculture, at Tacoma, Wash.

The new bruin has not been officially named, but he probably will be called the Lava bear, his known habitat being the lava beds of southeastern Oregon. The discovery was reported by Director Jewett of the United States biological survey. Mr. Jewett said that about a year ago a settler killed the first of these strange bears and brought the carcass to him. The animal had a shaggy coat very much like a grizzly bear and despite that it weighted only 40 pounds it was exceedingly ferocious.

Mr. Jewett determined that the animal was full grown and that its characteristics were distinctly different from those of any variety of bear known to him. He sent the carcass to C. Hart Merriam, former chief of the biological survey and now director of biological investigation for the Rockefeller institute. Mr. Merriam is the recognized world authority on bears.

Mr. Merriam confirmed the fact that the animal was different from any other known species but not much was made of the matter until two weeks ago when another animal of the same sort was killed by investigators of the biological survey. The experts' theory is that the Lava bear is a development from the grizzly in the district where it was found. Food is very scarce there and just as the huge Kodiak bear of Alaska is believed to be a grizzly developed to abnormal size by feeding on the salmon which abound in the rivers there it is thought the Lava bear may be a grizzly stunted through many generations of scant feeding.

Its characteristics, however, are so different from any known variety of bear that Doctor Merriam declares that it must certainly be classed as a new species.

—Kingston (NY) *Daily Freeman*, August 16, 1923.

Unfortunately, the lava bear (whether aberrant individuals, race, subspecies, or species) is now extinct. At least, I've not seen any recent sighting reports. Such a small grizzly was not without precedent—there is a race of smaller brown bears in the Gobi Desert of Mongolia. There was an expedition by taxidermist Robert Limbert to Idaho lava fields in search of a dwarf grizzly in the early 1920s, but without success. His photographic work and writings about those lava fields, however, led to the formation of Craters of the Moon National Monument and Preserve.

Elizabeth's Mystery
A Strange Creature Has Stirred Up Excitement Over There.

Elizabeth *Herald*: The section just above the borough is much worked up over stories of some kind of creature that is said to be roaming about the region bordering the river. It is supposed to be some kind of amphibious creature heretofore unknown in these parts. Persons who claim to have seen it say it is a quadruped about four feet long, with a head like that of a hog. It is blamed with having chased boys out of a corn field where they were working and of having put a number of cattle in a field in a state of terror. It comes up from the river and retires there after its excursions on land, and on being followed its path through the high wood was plainly apparent and in the soft mud it left tracks of two web feet and two like those made by the hoofs of an ordinary hog.

A youth in the vicinity who has been studying up natural history says it fits the description of an animal described therein as a water-hog. One theory is that it may have escaped from a show or zoological garden. It has not thus far done anything more than scare boys and cattle, no injury being reported from its depredations.

—Connellsville (PA) *Courier*, July 18, 1903.

Elizabeth is in Allegheny Co., Pennsylvania. And, yes, the astute reader is correct in surmising that "water-hog" is another name for the South American capybara, the world's largest living rodent. That is actually a very plausible candidate for this mystery animal. Capybaras were, and still are, common in menageries.

The Capture of a Strange Creature.

The Philadelphia *Record* of Thursday says: The strangest creature ever seen in these waters was captured in the Delaware river at Burlington yesterday morning by Charles Wooden and Charles Adams while they were fishing for shad. It is about six feet long, with a large head shaped like a bulldog's and an immense mouth furnished with two rows of sharp teeth. The head is attached to the body by a long, sinuous neck and the small and deep sunken eyes are protected by long lashes. The body, which gradually tapers to the tail, is covered by a short fine fur, and two short imperfectly formed legs, with webbed feet like those of a duck, are attached just below the neck. The tail is peculiarly formed, having four blades exactly like the screws of a propeller. The strange creature was captured with difficulty. It fought hard, and, uttering a noise that was half hiss, half bark, it seized an oar in its mouth and crunched it to splinters. A strange odor resembling musk was emitted. Repeated blows of a hatchet disabled the animal and enabled its capture.

—Denton (MD) *Journal*, April 20, 1889.

This is a pinniped of some sort. Even today, northern ocean seals will occasionally wander south during the winter and end up in the Delaware Bay.

Irish Monster
Farmer Shoots—and Misses—From Two Yards as Dromate Lake Ogopogo Chases Off Brother

Dublin, Aug. 29.—(CP Reuter)—The monster of Dromate Lake, County Monaghan, Eire, raised his ugly head above the waters Monday, ducked under again before the fire of a gun-toting local farmer, and put an end to all local fishing and bathing until somebody gets the jinx or he disappears into the realm of legend.

The monster "appeared" before local residents a short while ago but his activities went unreported amid the welter of war news. Monday, however, a 30-year-old farmer who was out in a boat on a monster-hunt with other men said he had a shot at him from only two yards away, but apparently missed.

"I saw it over the edge of the boat and fired," he said. "It gave a splash and raised a big wave in the water, then disappeared." The farmer, who has lived all his life near the lake, said the monster's body was not fully visible but appeared to be about seven feet long, had two arms with webbed or closed feet and a tail about 18 inches long and six inches broad.

The hunt continues. Meanwhile, local residents have refused to go bathing and boat-fishing has been abandoned for cautious rod-and-line operations from the banks of the lake.

—Lethbridge (Alberta, Canada) *Herald*, August 29, 1944.

Here's another area that could use a more comprehensive evaluation. Yes, there are general UK cryptozoological and Fortean books that note the occasional Irish lake monster, but I suspect there's a great deal more to be examined—even if they do all turn out to be big eels. Interestingly, as John Kirk noted (*In the Domain of the Lake Monsters*, 1998), there may be a connection to an otter-like cryptid, which certainly isn't outside the description given in this newspaper account.

A New "Sea Monster" Appears
British Columbia Man Tells of Seeing it at West's Lake.
B. C. MacCallum in the Vancouver Province.

Another "demon monster" has made its appearance in British Columbia, and the description tallies in many respects with that credited the Okanagan Lake "mystery creature." The latest addition to British Columbia's water curios has been discovered in West's Lake, a very deep body of water about two miles wide and four miles long, on Nelson Island in Jervis Inlet.

For more than half a century the residents have discussed among themselves, in tones of great awe, tales of a huge "demon" living in this lake.

They have called it the "Shelillican," and by that name it is known by the natives up and down the coast.

The only white man who claims to have even seen the "Shelillican," is Mr. John West, who settled on the foreshore of West's Lake thirty-five years ago. The first winter he was there he heard many stories of the monster from different Indians. The most complete tale he heard was from Big August, an Indian who, while staying at his house during a very bad storm, told of the activities of the great demon and gave a description of it.

It was five years after he first settled on the lake shore that Mr. West first saw what he thought might have been the "Shelillican." He was standing on the shore when at some distance he sighted what he at first thought was a canoe, but as he watched the object it suddenly disappeared. Though he continued to hear the Indians talk of the "Shelillican" he did not see anything unusual in the lake for thirty years.

Not long ago Mr. West was proceeding towards his home in a canoe. It was a beautiful day and the war sun was on his back as he paddled along. He had not gone far when a black object in the lake ahead attracted his attention.

When less than one hundred yards away Mr. West saw the "Shelillican" for there was no doubt that he was looking at the much-talked-of demon of the lake.

"In many ways it resembled a turtle," said Mr. West in describing the "Shelillican." "There were two parts showing above the water, one the head and the other a portion of the back. The head was about the size of that of a calf, the cheeks were yellow and its features were like those of a monkey. The visible portion of the back was about six feet long and shaped like that of a deer. While I was watching this peculiar monster it turned its head. As it did so the sunlight reflected from its eyes. Then it saw me and sank beneath the water with scarcely a ripple."

Mr. West was at a loss to account for the presence of the "Shelillican," but it is his belief that whatever it is its years are many and its species is that of a bygone age.

The lake in which the "Shelillican" lives is about two miles wide and four lives is about two miles wide by four the bottom has never been reached except near the shore.

The general belief prevalent among the Indians is that the "Shelillican" lives at a great depth and comes to the surface only in particularly warm weather. It is thought that it lives on herbaceous matter and the deep mud on the lake bottom near the shores is covered with a thick matting of a long grassy growth.

—Kansas City (MO) *Star*, Dec. 18, 1925.

This is one of the myriad accounts of aquatic mystery animals in the lakes and off-shore waters of British Columbia. West Lake, according to VancouverIsland.com, is difficult to reach, being almost entirely surrounded by private land. The only public access requires a "2-km bushwack on the west side of Vanguard Bay off Jervis Inlet—over second-growth forest up and down a 600-ft ridge with no trail." The lake has approximately 593 hectares of surface, or about 2.3 square miles. There's no mention of lake monsters on the site.

The best thing about cryptozoology is that investigations are not limited to far-off rainforests, deserts, or mountain ranges. Legitimate (and interesting) investigations can be close to home, often with mystery animals without wide public recognition.

Sometimes this research requires field work, other times it may be historical search and analysis. But as piece after piece is acquired and scrutinized, we begin to grasp a picture that eventually becomes the starting point for a more rigorous methodology. As we become more critical, we recognize and pursue better evidence. With better evidence, conclusions have greater support. We have a responsibility within cryptozoology to take nothing at face value, yet consider reports carefully, rather than dismissing them *a priori* or based on a poor understanding of biological principles. For those readers planning to investigate little-known cryptids in their local region, feel free to contact me through Strangeark.com. I would enjoy hearing about the project and may be able to suggest resources and methodologies for consideration.

About the Authors

Chad Arment lives in Lancaster County, Pennsylvania. His cryptozoological interests range from herpetological mystery fauna to methodology, historical research to lesser-known cryptids. He runs StrangeArk.com and published *Cryptozoology: Science & Speculation*.

Matt Bille is a science writer in Colorado Springs, CO. A former Air Force officer, he is a defense consultant when not pursuing his real interests. He has written one book on space history (*The First Space Race*, Texas A&M, 2004) and two on discoveries in zoology and crypotozoology (*Rumors of Existence*, 1995, and *Shadows of Existence*, 2006, both from Hancock House). Contact information and excerpts from his books, as well as his weblog on science and technology, can be found at www.mattwriter.com.

Gary Mangiacopra's passion for cryptozoology has resulted in numerous papers researched and published on almost every topic imaginable. Gary is an internationally recognized expert in such fields as cryptofiction and the ecology of sea serpents and lake cryptids. Gary received his M. S. in Biology working with Dwight Smith. At their first meeting, Gary approached Dwight about working on a thesis about lake cryptids. Dwight suggested that Gary use an ecological approach and the rest—as they say—is history. The two have worked on many projects and hope to continue their work on the use/application of ecological concepts, theories, and principles in the study of cryptozoology.

Barton M. Nunnelly, a native of Western Kentucky, has investigated all things cryptozoological for over 20 years. Located in Henderson, KY, he is the co-founder of www.kentuckybigfoot.com and has an avid interest in all unclassified animal reports, particularly those from the Bluegrass state, where he claims personal encounters with several different cryptids including Bigfoot, Black Panthers, aquatic unknowns and a Thunderbird.

Jerry A. Padilla is a historian-researcher and section editor of *The Taos News* Spanish and Regional sections, *El Crepúsculo* and *Valle Vista*. *The Taos News* is a northern New Mexico weekly based in Taos, New Mexico. Padilla has had a life long interest in the multicultural and historical traditions of New Mexico and the Southwest. With a special interest in wildlife, cryptozoology, and the outdoors, he has previously researched and published articles in Spanish for *El Crepúsculo* about cryptozoology, rare and endangered animal and fish species and special features on New Mexico and Mexican wildlife.

French cryptozoologist Michel Raynal was born in 1955, and worked for 10 years as a biochemist after his biology studies at the University. He became interested in cryptozoology in 1975, while reading Bernard Heuvelmans's books. Raynal then began a huge correspondance (two archive boxes) with Heuvelmans, who encouraged him to do personal research on ill-known or completely new cryptozoological subjects, such as the mysterious wingless bird from Hiva-Oa.

Dwight G. Smith received his B.S. in biology from Elizabethtown College in Lancaster County, Pennsylvania. He attended Brigham Young University in Provo, Utah, and received his Ph.D. His studies focused on ecology, population ecology, and raptor ecology of the Great Basin Desert. Since that time, he has conducted research in Alaska, Siberia, South Africa, and South America, primarily on population ecology of raptors and their prey. Dwight has published 15 books and about 500 papers on his work. Since 1970, Dwight has taught biology at Southern Connecticut State University in New Haven, Connecticut, where he is currently professor and Chair of the Biology Department.

Nick Sucik is currently an Anthropology student attending Northern Arizona University. Originally from Minnesota, he served in the Marine Corps as an infantryman, stationed in Hawaii before moving to Arizona. There, he worked as a legal liason for Navajo families living on land partitioned to the Hopi Tribe. Nick has conducted field research on lesser-known cryptids throughout the southwestern United States and also in Ireland, Italy, and Serbia. He operates MysteryAnimalsOf Ireland.com, along with working with two colleagues in Europe on a site devoted exclusively to European cryptozoology at EuropaCZ.com.

Also available from Coachwhip Publications

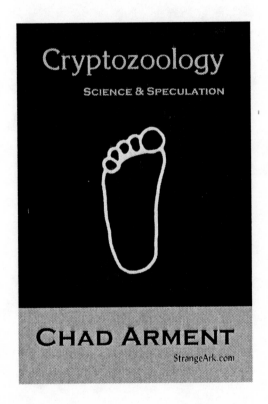

Cryptozoology: Science & Speculation
Chad Arment
ISBN 1-930585-15-2

CoachwhipBooks.com

Printed in the United States
56105LVS00003BA/203